CW01269847

Palgrave Series in Indian Ocean World Studies

General Editor
Gwyn Campbell, Indian Ocean World Centre, McGill University

Advisory Board
Philippe Beaujard, EHESS, CNRS, CEMAF, France
William Gervase Clarence-Smith, The School of Oriental and African Studies, University of London
Masashi Haneda, IASA, University of Tokyo
Michael Pearson, University of New South Wales
Anthony Reid, Australian National University
Abdul Sheriff, Zanzibar Indian Ocean Research Institute
James Francis Warren, Murdoch University

The Palgrave Series in Indian Ocean World Studies is the first series dedicated to the study of the Indian Ocean World from early times to the present day. It incorporates, and contributes to, key debates in a wide array of disciplines, including history, environmental studies, anthropology, archaeology, sociology, political science, geography, economics, law, and labor and gender studies. Moving beyond the restrictions imposed by Eurocentric time frames and national and regional studies analyses, this fundamentally interdisciplinary series is committed to exploring new paradigms with which to interpret past events, particularly those that are influenced by human-environment interaction. In this way, it provides readers with compelling new insights into areas from labor relations and migration to diplomacy and trade.

Starvation and the State: Famine, Slavery, and Power in Sudan, 1883–1956
  Steven Serels

Sailors, Slaves, and Immigrants: Bondage in the Indian Ocean World, 1750–1914
  Alessandro Stanziani

The Making of an Indian Ocean World-Economy, 1250–1650: Princes, Paddy Fields, and Bazaars
  Ravi Palat

The Portuguese in the Creole Indian Ocean: Essays in Historical Cosmopolitanism
  Fernando Rosa

Trade, Circulation, and Flow in the Indian Ocean World
  Edited By Michael Pearson

Histories of Medicine and Healing in the Indian Ocean World: The Medieval and Early Modern Period, Volume One
  Edited By Anna Winterbottom and Facil Tesfaye

Histories of Medicine and Healing in the Indian Ocean World: The Modern Period, Volume Two
  Edited By Anna Winterbottom and Facil Tesfaye

Histories of Medicine and Healing in the Indian Ocean World

The Medieval and Early Modern Period

Volume One

Edited by
Anna Winterbottom and Facil Tesfaye

palgrave
macmillan

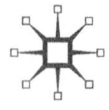

HISTORIES OF MEDICINE AND HEALING IN THE INDIAN OCEAN WORLD
Selection and editorial content © Anna Winterbottom and Facil Tesfaye 2016
Individual chapters © their respective contributors 2016

All rights reserved. No reproduction, copy or transmission of this publication may be made without written permission. No portion of this publication may be reproduced, copied or transmitted save with written permission. In accordance with the provisions of the Copyright, Designs and Patents Act 1988, or under the terms of any licence permitting limited copying issued by the Copyright Licensing Agency, Saffron House, 6-10 Kirby Street, London EC1N 8TS.

Any person who does any unauthorized act in relation to this publication may be liable to criminal prosecution and civil claims for damages.

First published 2016 by
PALGRAVE MACMILLAN

The authors have asserted their rights to be identified as the authors of this work in accordance with the Copyright, Designs and Patents Act 1988.

Palgrave Macmillan in the UK is an imprint of Macmillan Publishers Limited, registered in England, company number 785998, of Houndmills, Basingstoke, Hampshire, RG21 6XS.

Palgrave Macmillan in the US is a division of Nature America, Inc., One New York Plaza, Suite 4500, New York, NY 10004-1562.

Palgrave Macmillan is the global academic imprint of the above companies and has companies and representatives throughout the world.

Hardback ISBN: 978–1–137–56760–4
E-PUB ISBN: 978–1–137–56755–0
E-PDF ISBN: 978–1–137–56757–4
DOI: 10.1057/9781137567574

Distribution in the UK, Europe and the rest of the world is by Palgrave Macmillan®, a division of Macmillan Publishers Limited, registered in England, company number 785998, of Houndmills, Basingstoke, Hampshire RG21 6XS.

Library of Congress Cataloging-in-Publication Data

Winterbottom, Anna, 1979–
   Histories of medicine and healing in the Indian Ocean world / Anna Winterbottom and Facil Tesfaye.
       volumes cm.—(Palgrave series in Indian Ocean world studies)
   Includes bibliographical references and index.
   Contents: volume 1. The medieval and early modern period—volume 2. The modern period
       ISBN 978–1–137–56760–4 (hardback : alk. paper)
   1. Medicine—Indian Ocean Region—History. 2. Medicine—Indian Ocean Region. 3. Medicine, Medieval—Indian Ocean Region. 4. Traditional medicine—Indian Ocean Region—History. 5. Traditional medicine—Indian Ocean Region. 6. Healing—Indian Ocean Region. 7. Indian Ocean Region—History. I. Tesfaye, Facil. II. Title.

R605.W46 2015
615.8'8091824—dc23   2015018001

A catalogue record for the book is available from the British Library.

These volumes are dedicated to medical workers around the world who risk their lives to help others in situations of conflict.

Contents

List of Figures — ix

Acknowledgments — xi

Introduction — 1
Anna Winterbottom and Facil Tesfaye

1 An Insight into al-Razi's Extraordinary Theoretical and Practical Contributions for Developing Arthrology — 37
   Mahmud Angrini

2 Exchanges and Transformations in Gendered Medicine on the Maritime Silk Road: Evidence from the Thirteenth-Century Java Sea Wreck — 63
   Amanda Respess and Lisa C. Niziolek

3 Saints, Serpents, and Terrifying Goddesses: Fertility Culture on the Malabar Coast (c. 1500–1800) — 99
   P. K. Yasser Arafath

4 The Circulation of Medical Knowledge through Tamil Manuscripts in Early Modern Paris, Halle, Copenhagen, and London — 125
   S. Jeyaseela Stephen

5 Medicine, Money, and the Making of the East India Company State: William Roxburgh in Madras, c. 1790 — 151
   Minakshi Menon

Bibliography — 179

Notes on Contributors — 195

Index — 197

Figures

1.1	The first page of the manuscript. Left: the copy in the library of Cambridge University (MS 3516); right: the copy in the Malek Library in Tehran, Iran (MS 4442/i)	40
1.2	The first page of the second copy in the library of Malek Library in Tehran, Iran (MS 4573/8i)	41
1.3	The approximate percentages for the contents of the treatise	44
1.4	The theory that al-Razi depended on to explain arthralgia	48
1.5	Great saphenous vein	55
2.1	Location of the *Java Sea Wreck*	64
2.2	Export routes and regions where Chinese ceramics were imported during the Song Dynasty	65
2.3	Sources of ivory and routes for the ivory trade during the Song Dynasty	69
2.4	Elephant tusks from the *Java Sea Wreck*	70
2.5	A block of resin from the *Java Sea Wreck*	71
2.6	*Kundika* from the *Java Sea Wreck*	74
2.7	*Kendi* from the *Java Sea Wreck*	75
2.8	Ambrosia bottle from the *Java Sea Wreck*	76
2.9	Gourd-shaped *kendi* from the *Java Sea Wreck*	77
2.10	Gourd-shaped qingbai ewer from the *Java Sea Wreck*	78
2.11	Qingbai-style covered boxes from the *Java Sea Wreck*	79
2.12	Covered box lid with pomegranate design from the *Java Sea Wreck*	80
2.13	Covered box lid with boteh design from the *Java Sea Wreck*	80
2.14	Qingbai-style covered box lid with basket weave design from the *Java Sea Wreck*	81
2.15	Qingbai-style ribbed, melon-shaped box from the *Java Sea Wreck*	81
2.16	Qingbai-style covered box base with sculptural couple from the *Java Sea Wreck*	82
2.17	Bowl with design similar to the Taiji symbol in the center from the *Java Sea Wreck*	86

2.18	Qingbai-style covered box base with molded phallic design (highlighted in black) from the *Java Sea Wreck*	87
2.19	Qingbai-style covered box base with molded flower and inscription (highlighted in black) from the *Java Sea Wreck*	91
2.20	Qingbai-style covered box base with inscription (highlighted in black) from the *Java Sea Wreck*	92
5.1	Herbarium sheet, *Nerium tinctorium*, labelled in Roxburgh's hand	172

Acknowledgments

We would like to thank the Indian Ocean World Centre, directed by Professor Gwyn Campbell, for hosting the conference "Histories of Medicine in the Indian Ocean World" from which the process of writing this book began. We are also very grateful to Professor Nicholas Dew and the SSHRC Situating Science project, and to Professor David Wright and the Institute for Health and Social Policy for financial and logistical support. We are grateful to Professor Howard Philips for joining us as the keynote speaker. We would like to thank Erin Bell, Lori Callaghan, Omri Bassewitch Frenkel, Peter Hyde, Lorna Mungur, Caroline Seagle, and Dr. Hideaki Suzuki for their help in organizing the conference and Professors Rachel Berger, Laurence Monnais, Andrew Ivaska, and Jon Soske for chairing sessions and commenting on papers. Many thanks are also due to Christopher Lyons for providing a tour of the Osler library. We are grateful to Melissa Kawaguchi for preparing the index.

Introduction

Anna Winterbottom and Facil Tesfaye

The monsoon that binds the Indian Ocean together is an agent of both connectivity and chaos,[1] as its currents carry not just people and their goods but also diseases.[2] Healing follows disease. The authors of the chapters collected in this book approach their studies from different vantage points, spanning more than a millennium. However, they all address the topics of disease, medicine, and healing within and across the geographic and conceptual space constituting the "Indian Ocean world" (IOW).

In his keynote address at the conference from which this book emerged, Howard Philips referred to the "Swahilian swap of pathogens."[3] As a physical event, the monsoon creates an environment for disease, as the mosquitos that breed in the pools the rains leave behind spread dengue, malaria, and *chikungunya*. In his pioneering account of the Indian Ocean as a "disease zone," David Arnold observed that contagious diseases, including bubonic plague, and perhaps smallpox and cholera, spread from centers of dense population such as India and China, a process that accelerated with the great population movements of the industrial age.[4] New arrivals in the region also brought disease to their host environment. Michael Pearson notes in his afterword to this book, that historically, this is perhaps best expressed in the name "firangi" or "parangi," which derives from "foreigner" but also denotes syphilis.[5]

Healing therapies sometimes reference the monsoon specifically. One could mention, for instance, the Kerala Panchakarma therapy with its focus on rejuvenation during the rains—that now attracts tourists from across the Indian Ocean region—which is actually matched to the seasonal monsoon. Historically, travelers also voyaged across the Indian Ocean in search of medical treatment and religious healing as well as profitable items of materia medica. Reading the list of ingredients specified in almost any premodern pharmacopoeia from the region is very revealing. The reader is taken on a journey around the IOW as he or she learns about Egyptian opium, Socotra aloes, the ambergris of Azania, Syrian sumac, Armenian bole, Persian sweetmeats, Indian aloeswood, and the cinnamon of Lanka, to Southeast Asian spices and the rhubarb, celandine, and ginger of China.[6] For an alchemic prescription one might travel to Yemen for talc, Istahr for iron oxide, Zarawant for borax, or Kīrman for malachite.[7]

The concept of an Indian Ocean "world" is still relatively new. In bringing together the chapters in this book, we seek, in various ways, to test the claim that the region might be regarded as a unit of historical analysis. While there has been no previous study of medicine and healing in this region, some scholars have touched on elements of our subject from various angles. Therefore, before moving on to the presentation of the chapters of this book, we will review some of the most important works that have briefly mentioned medicine in the IOW.

The initial attention to the Indian Ocean region came from economic historians inspired by Ferdinand Braudel's account of the Mediterranean.[8] Medicines and objects used in medical practice were, of course, also items of trade. Periplus, produced in around 40–100 CE and usually regarded as the earliest account of the commercial landscape of the Indian Ocean, already noted the exchange of goods such as myrrh and frankincense, which, like other aromatics, also have medical uses. Furthermore, materia medica overlapped with foodstuff such as sugar, cinnamon, pepper, nutmeg, and other spices traded across the region that were also used for medical purposes. Animal parts and products, including ivory, horn, ambergris, the famous bezoar stones, and even cowrie shells that doubled as currency, were in demand for their medical uses in the IOW. The works of Auguste Toussaint,[9] K. N. Chaudhuri,[10] and others who followed in their footsteps convey some sense of the importance of these items in trade and tribute.[11] Economic atlases, including the modern editions of Irfan Habib and Brice,[12] provide a visual account of the specialization of particular regions in producing medicinal plants for trade. Although these economic histories of the IOW mention the trade in medical products, none of these works use the terms "medicine" or "healing" explicitly or provide information about how materia medica were used and conceptualized in the different regions around the Indian Ocean.

Studying medical practice by religion, language, or cultural grouping is one way in which historians have approached transregional exchanges. Buddhism has been recognized for its role in disseminating medical teachings between South, Southeast, and East Asia from the fifth century CE onward.[13] Some authors have noted that medicines joined textiles and religions in being transported via the "silk road." A recent study of Japanese medicine has even imagined a maritime "pharmaceutical silk road," established from the eighth century and manifest in writings including those of the fourteenth-century Buddhist monk Kajiwara Shōzen. The latter had a significant number of ingredients, listed in his pharmacopeia, which he had received from all over the IOW (called *Nanban* or the Southern Barbarian Region) as well as from China and the Ryukyu islands.[14] The exchanges were not only limited to materia medica, but also included medical books, students, and doctors. Doctors travelled for both medical training and diplomatic reasons, thus highlighting their importance in establishing the connections across what has been referred to as the "East Asian Mediterranean."[15] While the spread of Hinduism has been more diffuse and subtle than that of Buddhism, the "Hinduization" of much of Southeast Asia, beginning gradually from the first century and being consolidated between the seventh and tenth centuries, saw the spread of medicinal plants along with scripts and deities.[16] The turn toward the West from the sixteenth century onward, where Hindus increasingly acted as subcolonists,[17] also saw the transplantation flow westward across the Indian Ocean. This is perhaps best evidenced by the appearance of the neem tree (*Azadirachta indica*) as an important source of multiple items of materia medica in Eastern Africa, Saudi Arabia, Iran, and finally West Africa and the Americas from the nineteenth century onward.[18]

The rise of Islam in the seventh to eighth centuries CE coincided with the period that Chaudhuri sees as the maturing of the Indian Ocean as an interconnected zone for the distribution of goods. Historians including Sanjay Subrahmanyam have stressed the heterodox nature of Islam in the region.[19] Nevertheless, the sharing of key Islamic beliefs has often been conceptualized as a uniting force within the Indian Ocean. Ho opens his account of the Hadrami society with a mariner's poem, "stringing along ports like prayer beads," which attaches the name of a saint to each port.[20] If the hinterlands remained unknown, the Friday mosque provided a familiar point of reference for Muslim traders in many of the entrepôts around the Indian Ocean.[21] This also meant that the influence of Islamic concepts of medicine quickly spread along the routes taken by traders, pilgrims, and proselytizers. Protocols of Islamic medicine are known to have influenced the theoretical developments in Chinese medicine during the Tang and Song periods and, by extension, influenced medieval Japanese concepts.[22] Further West, *unani tibb* (Graeco-Islamic medicine) has traditionally received attention in histories of medicine as the bearer of the classical tradition during the European Middle Ages and for its role in disseminating humoral theory.[23] Sufis considered the graves of saints to be places where "mobile persons and mobile texts meet." These were also sites where disagreements about the relationship between living and dead were played out.[24] Nile Green describes a pilgrimage to saints' graves, which became an established practice by at least the tenth century, as "a key space-making institution introduced at an early period from a wider Islamic system of settlement and acculturation, especially in erstwhile frontier territories of Africa and Southeast Asia as well as India."[25] Tombs were also considered as sites of healing since the "blessed power" of the saint (*barakat*) could include the performance of cures as well as the resolution of worldly problems and disputes. This practice was, in some cases, accomplished by women and other people who were outside the official religious sphere.[26] In the case of larger shrines, like that of Imam Reza in Iran, hospitals were set up to provide for the needs of the pilgrims who required medical attention.[27] Several authors have made the general point that the spread of ideas, including medical ideas, and the expansion of Islam not only acted as a force for social cohesion but also created "zones of political tension"[28] or "zones of moral competition."[29] It is, in fact, well known that Sufi shrines have been targets for adherents of certain forms of Islam and, in some cases, for political authorities who consider them a threat. As evidenced by such examples, the experience of migration can act to strengthen localized attachments of class, caste, or tribe, rather than dissolving them. Moreover, communities that retained strong networks across the Indian Ocean, such as Bohra or Khoja, held widely differing social practices across and beyond the region.[30]

Scholars have, however, also noted the existence of similar healing practices along the contours of the IOW. The growing recognition of the presence of people of African descent throughout the IOW has led some scholars to focus on African healing beyond the continent. They argue that, despite their ambiguous relationship with forms of healing based on textual scholarship, African spirit cults that often spread along with Islam combined spirit possession with healing. Edward Alpers, who belongs to this group of scholars, notes that the "important innovations and retentions in music, song and dance, spirit possession and healing, medical pluralism and popular religion,…are linked and can be compared with the situation in the Americas and the Caribbean islands."[31] Helen Basu, who explores the intertwining of Sufi Islam and African-derived cosmologies and performance styles, also indicates that the symbol of the drum (*ngoma*) in East Africa is recalled by the ritual performances of Sidi communities in Gujarat. Meanwhile, she also pinpoints the problems that could come along with a straightforward definition of Sidis as a "diaspora" as well as with essentializing their ritual practices as having an African origin. She argues for the need to allow for transformation in the Indian Ocean context and dialogue with both Hindu and Muslim concepts of spirit possession and healing.[32] Perhaps the best-known example of the migration of a recognizable spirit cult associated with healing is the women's medicine *zar* cult practiced in North Africa, the Horn of Africa, and the Middle East, and that also resembles the West African *bori*.[33] As with this example, the exchange of healing techniques was, of course, not limited to the boundaries of the IOW. In fact, other practices that may have originated from elsewhere were also drawn into the circulation within the region. While spirit possession cults have attracted the attention of scholars, only recently have historians of the environment begun to pay attention to the transplantation of medicinal plants such as the baobab tree from Africa to other regions around the Indian Ocean.[34]

Like shared religious practices, lingua franca also facilitated the translation and distribution of medical information across the region. The pharmacological tradition in the Arabic language not only built on the work of Dioscorides, but also incorporated information from translations of Sanskrit works. The Kitāb al-ṣaidla fī al-ṭibb (Book on the Pharmacopoeia of Medicine) of al-Biruni—or Alberuni, as the tenth-century Persian polymath was known in the West—is a case in point. The book uses an array of sources from across the region and serves as a reference document. For instance, it gives drug synonyms in Syriac, Sankrit, Persian, Greek, Baluchi, Afghan, Kurdi, Indian dialects, and other languages.[35] While the "golden age" of Arabic pharmacology was traditionally considered to have ended in the eleventh century, many later examples of innovation exist. For instance, Leigh Chipman's in-depth study of pharmacological texts from late Mamlūk Cairo produced among the Jewish community demonstrates a high degree of cosmopolitanism in the sourcing of ingredients from around the Indian Ocean region, including China.[36] The comparable work of the twelfth-century Nestorian Christian of Baghdad, Ibn at-Tilmīd, could be mentioned as another example.[37] Furthermore, sources indicate that Persian medical literature found audiences from Cairo to Delhi,[38] and that this process was aided by the frequent migrations of doctors, which Cyril Elgood refers to as an "exodus" during the Safavid period.[39] Seema Alavi and Guy Attewell's recent studies of *unani tibb* in India both highlight the ongoing interactions between Persian and Urdu medical literature.[40]

Institutional histories of the practice of medicine in the region also demonstrate exchanges across the IOW. An example is the recent collection of essays pertaining to hospitals in Iran and India edited by Fabrizio Speciale.[41] While these cover the period after 1500, the institution of the hospital in the region is far older. The edicts of the Buddhist Emperor Aśoka (c. 259–222 BC) famously ordered the construction of institutions of healing, although their effects are unknown. The remains of the Mihintale hospital in Sri Lanka probably date back to the reign of either King Sena II (851–885 AD) or King Mahinda IV (956–972 AD).[42] As demonstrated by the Persian medicine jars and Chinese ceramics excavated at Mihintale and other sites in Sri Lanka, the hospitals' supply chain was cosmopolitan. The idea of a dedicated site to heal the sick was widespread by the beginning of the second century CE. Islamic hospitals were constructed from medieval Spain[43] to the Delhi Sultanate. Some of the most eminent physicians of the age divided their time between court and hospital: Abu Bakr Muhammed ibn Zakariyā Rāzī, known as al-Razi, or Rhazes in the West (c. 859–925 CE) of Rayy—whose work is discussed by Mahmud Angrini in this book—worked at the Baghdad hospital and Abū'l Ḥasan 'Alī ibn al-Nafīs presided over the hospital founded in Cairo by Qalā'ūn, the Baḥrī Mamlūk Sultan of Egypt.

Bathing had been an integral part of hospital practice in the region from an early stage. Sources indicate that the "sarcophagus-like" immersion tanks found among the ruins of the ancient Sri Lankan hospitals were used for the application of medicinal herbs. In the Muslim world, the *ḥammām* was regarded as a place of healing, although it was also sometimes considered as the abode of disease-causing djinns.[44] *Ḥammāms* were often found within the same complex as hospitals, along with the caravansary, serving the poor or the itinerant who did not have access to the private luxuries of medicine, food, and cleanliness.[45] These examples form part of a wider international culture of healing through bathing: spas historically received state support in several parts of the world, and are being reinvented as "alternative" therapies in many parts of the contemporary world.[46] Spa treatments sometimes, although not invariably, are directly connected to the ocean, as they involve the use of seawater, or "thalassotherapy," and are often associated with the application of minerals, seaweed, or algae as medicines. Massage, often performed within the setting of the spa, is accepted in the modern world as both a mainstream therapy in the form of chiropractic manipulation, and as complementary medicine.[47] Both baths and massages are generally associated with the pleasurable face of healing as well as sometimes with sexual activity, one reason for their sometimes-controversial status within medical establishments.

In both the Mughal and Safavid empires, a hospital complex could also include a mosque and madrasa. Although the madrasas' curriculums would sometimes have included works such as the *Qānūn* or "Canon of Medicine', by Abū 'Alī al-Ḥusayn ibn 'Abd Allāh ibn Sīnā, known as Ibn Sina or in the West as Avicenna, they were not generally specialist medical schools. The hospital was not directly affected by religion, since Hindu doctors were employed by Muslim-endowed hospitals in India;[48] while in the Qajar period in Iran, hospitals offering allopathic medicine were often endowed as through the *waqf*. State support for hospitals in India began with the Mughal Emperor Jahangir's command that they were to be funded through the ḫāliṣa, or the lands directly managed by the imperial government.[49] Studying dispensaries provides a smaller-scale institutional history of medicine, one that often intersects with family histories, as businesses—such as the Hamdard dispensary—were passed down through generations.[50]

The European colonial presence in the Indian Ocean from the early modern period onward has provided one of the most obvious entries for scholars into the subject of medical exchanges in the region. Health was, in fact, both a practical and an ideological concern for the outsiders who sought to establish themselves in the region from the fifteenth century onward. As Michael Pearson notes in the afterword of this book, his own work[51] examines medical contacts between Portugal and India in the larger framework of Indian Ocean history. The medical aspect of the Portuguese quest for "Christians and spices" is discussed by Ines Županov[52] and, in previous studies, by Cristiana Bastos.[53] Natural and medical knowledge in the empire of the Dutch East India Company (VOC) has recently received attention from Harold Cook,[54] and Kapil Raj and Emma Spary have shown how the French empire's taste for natural history grew out of the study of exotic medical plants.[55] The British Empire has also been the focus of many studies of the social, cultural, and political aspects of disease and medicine. The concept of the colonial body became particularly central to the research of the subaltern studies collective.[56] While the aims of colonial states and missionaries were sometimes at odds, the purveyor of medicine on the ground was often the mission doctor. The latter became an emblematic figure in the later colonial period when Europeans claimed to heal the physical and moral sickness they perceived in the tropics, through a combination of "Christian conviction, imperial mission, and science."[57] Despite the numerous studies of colonial and missionary medicine, the focus has tended to be on the metropolitan policies of the particular colonizing country rather than engaging with the ways in which imperial medical policies interacted with one another within a specific location.[58]

One consequence of the colonial period was the emergence of a conceptual division between "Western" and "non-Western" medicine. The exact time when this perceived division emerged is debatable. Braudel argued that in the eighteenth century there was a shattering, in both China and Europe, of "a biological *ancien regime*, a set of restrictions, obstacles, structures, proportions, and numerical relationships that had hitherto been the norm."[59] The decline of humoral theory and the application of the chemical revolution to drug discovery (after its initial divorce from pharmacology) after around the mid-nineteenth century might also be regarded as turning points in the emergence of modern allopathy and biomedicine. How far any true "divergence" took place between the medical thought and practices of Europe and America and other parts of the world, has been seriously questioned by a number of recent studies, which have demonstrated that the reach of colonial medicine on the ground continued to be limited and its application shaped by local factors.[60] "Western" medicine was not the only medical tradition to profess universal applicability, as Alavi demonstrates in her discussion of Persian medical encyclopedias.[61] Nor was allopathic medicine the only tradition to adapt itself to new media and approaches, as Attewell notes in the context of Urdu-language medical journals[62] and as Cochran observes in his study of the commercialization of Chinese pharmacy.[63] Nonetheless, by the twentieth century, apologists for colonial regimes felt able to list "Western medicine," along with railways and canals, as a boon that they claimed to have bestowed on the colonies.[64]

Despite the increasingly disparaging approach of most colonial regimes toward practices that were considered to lie outside the scope of evidence-based medicine, the colonial period also saw the beginnings of an international market for "alternative" healing. These forms of healing notably included yoga and meditation, and they were aided by an interest in their spiritual as well as medical benefits.[65] While these practices can be, and often are, divorced from the theoretical framework within which they were originally located, the global demand for different approaches to healing has continued to grow since. Often this demand has arisen in response to the perceived disadvantages of allopathic medicine, including the negative side effects of drugs and the lack of a "person-centered" approach. In this respect, Ayurveda,[66] Chinese, Tibetan (*Sowa rigpa*),[67] and South African (*muti*) traditional medicines have attracted attention for their contemporary transnational nature as they are repackaged for a global clientele. Recent studies have explored the complexities of the encounters of these healing practices with biomedicine. These cover the incentives to present themselves as complete philosophical packages, with principles opposed to those of biomedicine[68] and the incentives to adopt clinical trials of drugs or to develop alternative methods of testing.[69]

A parallel, but distinct strand in the globalization of healing practices sees the migration of specific therapies, medicines, or rituals for the consumption of diasporic communities. A good example of this is the migration of lên đồng from Vietnam to the United States.[70] Doctors continue to be among the most mobile groups of professionals, their trajectories of migration often being determined by the correspondence between languages and systems of education created by earlier colonial contacts.[71] Relations between doctors and wider migrant communities have been the focus of some debate, especially around the controversial issue of female genital cutting, also known as female circumcision or female genital mutilation.[72] Patients also continue to travel internationally in search of specific treatments. The modern practice of "medical tourism," which is also discussed by Michael Pearson in his afterword to this book, takes place in all directions, motivated by a number of concerns, including a search for cheaper medical procedures, better funded or more sophisticated facilities, a more "caring" environment, and different cultural norms permissive of certain medical procedures.[73] As healing becomes increasingly internationalized, older connections, such as those that exist across the Indian Ocean, are not lost, although they are often transformed by new contexts. An example is the ongoing connection between South Asian and South African medicine.[74]

While both area studies and medical anthropology have provided important contributions on medicine in local and international contexts, few historical studies have engaged with the interconnection or disjuncture of healing practices across the Indian Ocean region. The foci of the studies presented here are on very specific moments in medicine, but they all situate healing within a regional context.

The contributions in this book are drawn from historians, archaeologists, anthropologists, geographers, area studies specialists, and health professionals and link economic, intellectual, cultural, and social histories. In doing so, they demonstrate the wide spectrum of areas of life impacted by the exchange and consumption of medical substances, practices, and ideas. The time period covered stretches from the ninth-century work of al-Razī, discussed by Mahmud Angrini, to the contemporary studies of Jonathan R. Walz and Julie Laplante. The chapters cover a geographical area stretching from the Chinese end of the "porcelain road" by Amanda Respess and Lisa C. Niziolek, to the meeting of the Indian and the Atlantic Oceans in Laplante's work on South Africa. In terms of the presence of outsiders in the region, Cristiana Bastos and Ana Roque discuss Portuguese imperial involvement in medicine that was spanning Goa and Mozambique. Karine Jansen discusses medicine in the Mascarenes in the period of French colonialism and its wake, whereas Yoshina Hurgobin's contribution makes a claim for a distinct medical culture among the plantation islands of the Indian Ocean, with a focus on the sugar colony of Mauritius. Minakshi Menon, Shirish Kavadi, and Anoushka Bhattacharyya all adopt an India-centered perspective on institutional innovations to deal with new plants, diseases, and public health challenges encountered in the British Empire, showing how the subcontinent acted as a center within the empire and the ocean. S. Jeyaseela Stephen examines the reception of Tamil medical texts by a range of European countries connected to the region through trade and missionary work. Rashed Chowdhury provides a perspective on the IOW from a power traditionally excluded from discussions of the region with his chapter on Russian diplomacy in Ethiopia.

A major part of healing normally consists in offering the patient an explanation for disease and its cure. The chapters touch on this theme in a number of ways. Many of the societies around the Indian Ocean shared some form of "humoral" theory, based on ideas of the balance of hot, cold, wet, and dry elements in the body and the environment and healing according to the theory of opposites. Mahmud Angrini's contribution, which opens the book, focuses on al-Razi: a figure who was influential in both transmitting and questioning the Galenic version of humoral theory. As Angrini shows, al-Razi's important contributions included the discovery of capillaries and the description of sciatica. In his study of one of al-Razi's lesser-known treatises, translated as "On Joint Pains" and preserved in manuscript in Tehran's Malek Library and in the University of Cambridge library, Angrini shows that al-Razi used the question of joint pains to evolve a humoral explanation for the affliction, as well as making pioneering anatomical observations. However, the bulk of the work was devoted to treatments, in which the ingredients and their administration were described in meticulous detail. al-Razi's treatise, the oldest surviving Arabic work on rheumatology, influenced physicians from Damascas to Samarqand to make their own studies Amanda Respess and Lisa C. Niziolek interrogate the meeting of Islamic humoral theory with Chinese yin-yang concepts of medicine and its implications for diagnostics in Song and Yuan periods in China. Like the doctrine of the four humors, the yin-yang in ancient Chinese medicine provided a broad organizing principle with which to categorize bodily and cosmological relationships. As these authors show, the reception of the work of Ibn Sina's Qānūn in China paralleled its dissemination in Western Europe between the twelfth and fourteenth centuries. Apart from providing a summary of Islamic medicine, Ibn Sina's work provided an influential exposition of Galen's version of the Hippocratic theory of the humors. It is notable that Isa Tarjaman, who Respess and Niziolek note was influential in formalizing the Islamic medicine in China, was a Nestorian Christian. As speakers of Syriac, Persian, and Arabic, the Nestorians provided an important channel for the dissemination of medical practices through the Near and Middle East as well as China.

As noted earlier, the objects on which healing is based (medicines or surgical instruments) and instructions for using them are often exchanged during trade. The way in which such objects are packaged often carries information about their significance, use, or meaning, information that enables such objects to successfully traverse cultures. Respess and Niziolek's contribution takes us into the hull of a Chinese ship wrecked off the coast of Java during the thirteenth century to examine a cargo that represents a cross-section of the healing objects that travelled with other items along what has been called the "Maritime Silk Road" or the "Porcelain Road." Many of these objects were concerned with the issue of fertility or reproductive medicine at the dawn of the emergence of *fuke*, or women's medicine, in China toward the close of the Song dynasty. In addition to goods made in China, including ceramics identified by Niziolek as having been made in the famous kilns in Jingdezhen in Jiangxi province, the collection from the Java sea wreck included Indian-style *kendis* and *kundikas* and elephant tusks that could have been derived originally from other Asian countries or from Africa. Ivory powder was a key component of women's medicine. Apart from having cosmetic applications, it was believed to correct sexual and energetic imbalances by regulating desire and sexual fluids. Also among the cargo were medicinal resins of Persian and Arab origin, which were often used in rituals connected with childbirth.

While al-Razi was dismissive of the power of talismans in medicine, they, nonetheless, formed a central part of the practice of his contemporaries, and magico-religious formulations from the Islamic world travelled far, as did medical texts. The forms of the healing objects discussed by Respess and Niziolek also provide unifying themes across religious traditions. The water vessels that were used in ritual ablutions within Hindu and Buddhist rituals began to be adopted for similar purposes within the Islamic world and eventually in China, where a modified form of the object was used in the brewing of particular medicines. In another example, the form of bottles designated in Indian art to contain the elixir of immortality was transferred to Chinese containers for fertility medicines. Ideas about gender could also be transferred through such objects, while being contextualized within familiar theories such as the yin-yang. Similarly, the shared symbolism of the melon and pomegranates promoted their similar uses within Chinese and Islamic pharmacological traditions, even when their effectiveness was explained using various different underlying theories. The increasingly "cosmopolitan" nature of the materia medica that was incorporated into Chinese medicine paralleled the shifts in Chinese diagnostic practices under the influence of works of Islamic medicine.

Yasser Arafath's chapter also shows how healing techniques could spill over the boundaries of religious beliefs. His chapter provides a view of the interactions between Hinduism, Islam, and Christianity as well as older Buddhist traditions through the lens of the shared elements of belief relating to fertility on the Malabar Coast. As he notes, healing traditions simultaneously reflect the plurality and the interdependence of cultural frameworks in the region. Like Respess and Niziolek, Arafath also explores the connections between fertility, bodily fluids, and the natural environment. Disease and cures were attributed to Sufi saints and holy men, gods, and especially goddesses, attached to a range of places including not only temples, dargahs, and churches, but also groves of trees (*kavu*), the abode of ancient snake deities. The same goddesses who were attributed the power of enabling fertility were also thought to cause smallpox. Like Respess and Niziolek, Arafath notes the centrality of objects, including clay pots, holy water, and written charms in rituals aimed at ensuring fertility. Sometimes these objects were combined in rituals such as consuming holy water from a ceramic bowl imprinted with Arabic letters arranged in pictographic tables together with numerals. The distribution of materia medica used in healing, such as the turmeric *prasadam* used at the Hindu Cheemeni Mundya temple at Kasargode, borrowed from similar practices within Buddhism, where healing is a function of the *vihara* or temple/monastery. The invocation of *nerchas*, or "divine personalities," is associated with international Islamic worship of saints but it also has Dravidian roots. The interaction between faiths was not always comfortable, as shown by the belief of Muslim women that they were possessed by *kafīr* (unbeliever) *djinns* or suffering the effects of the evil eye, and the widespread belief that the shape-shifting deities of the tribal communities could cause particular fertility problems. While the cause of disease was attributed to the malign influence of magic, diagnosis could be based on simple observations, which, at times, included observing patients' urine samples. Arafath also reflects on how healing interacted with perceptions of gender. While he notes that the worship of goddesses by women included potentially subversive rituals such as those performed at the "cock festival," he concludes that overall, fertility rituals served to entrench the control of men over the female body. Similarly, while the lower castes sometimes specialized in fertility and children's medicine and had access to the deities that inhabited *kavus*, prescriptions of ritual hygiene, nonetheless, served to maintain caste boundaries.

Different explanatory frameworks for disease often coexist with the overlapping use of medical substances or techniques. Several chapters in this book examine exchanges of medical knowledge. Whereas Respess and Niziolek focused on the transmission of medical knowledge along routes of trade and Arafath examines exchanges between communities of different religions, the chapters of Stephen and Menon both focus on the transfer of Southern

Indian botanical and medical knowledge to Europeans during the early period of colonial settlement in India. Stephen's chapter focuses on the northern section of the Tamil littoral known as the Coromandel Coast between 1700 and 1857, while Menon's focuses on the English East India Company (EIC) settlement of Madras, now Chennai, on the Coromandel Coast and the western hills of the Circars further inland.

Stephen examines the transmission of Tamil medical knowledge into works in French, English, Danish, and German as well as traces several palm leaf manuscripts that ultimately ended up in European collections. As Stephen points out, becoming acquainted with local remedies, such as those for snakebite, was essential for new arrivals on the Coromandel Coast. Therefore, both missionaries—Jesuit and Protestant—and representatives of the European trading companies were inspired to apply the study of Tamil, undertaken in the hope of spreading Christianity, to the search for medical knowledge. He demonstrates, through his archival research, not only the transfer of information from Tamil texts to European researchers, but also the acquisition of a text on rabies derived from the Tamil *Sillarai Kovai* by Louis Pasteur (1822–1865), the French biologist who is credited with inventing the rabies vaccine. Other texts that were investigated or partially translated into European languages included texts on medical theory and practice and human anatomy.

Stephen's contribution makes clear that, while language learning is important, the transfer of medical texts across cultures must be accompanied by close collaborative work or cultural immersion such as that undertaken by the Tranquebar missionaries Bartholomaus Zeigenbalg and Johann Ernst Grundler, since the meaning of such texts is not always transparent. While some studies have argued that European investigations of Asian medicine simply extricated the materia medica that were considered useful, Stephen shows that the work of Grundler goes much further in mirroring the layout of the original Tamil text, *Agastiyar Irandaayiram*, on which he based his work "The Tamil Physician," which was later included in the medical curriculum in Copenhagen. While some such texts successfully crossed into European medical teaching, other elements of Tamil medicine, such as the use of arsenic in the "Thanjavar pills" used against snakebite, were less readily accepted, given contemporary European concerns about the medical uses of poisons. The EIC surgeons Benjamin Hayne and Whitelaw Ainslie, while somewhat more reserved about the value of Tamil medical knowledge than Grundler, also facilitated the translation of Tamil medical texts into English. While noting the remarkable success in translating key concepts from Tamil to European languages, Stephen also notes the problems posed by the inclusion of poetic and secret elements in the original Tamil texts.

Both this chapter and Menon's note the official and unofficial support lent by the European trading companies and missionary groups to inquiries into local medicine made by their servants in India and beyond. However, both these chapters complicate Kuhn's concept of a scientific revolution involving an initial phase of European extraction of knowledge, followed by the dissemination of the new science from Europe to the colonies. Instead, they suggest a more dialectic, ongoing, and multidirectional exchange of knowledge.[75] The arguments of these chapters might sit rather better with the model advanced by Kapil Raj concerning the "circulation" of scientific knowledge during the period of European colonialism in South Asia.[76] However, both authors also ask what sort of knowledge circulated and why. Stephen discusses the barriers to the circulation of certain information, including the reluctance of Tamil physicians to disclose medical secrets, the particular conventions of Tamil medical manuscripts, or the prejudices of Europeans against the medical practices of other cultures. Menon's chapter argues for a closer look at the particular circumstances of those involved in the making and transmission of knowledge. Her contribution supports the argument made by Harold Cook, who notes that the interests of early modern European traders and naturalists converged around key concepts such as "credit."[77] According to this view, certain types of knowledge that were considered marketable and verifiable were more likely to travel than others. Identifying and exploiting transportable knowledge was thus a profitable pursuit. Menon follows Appadurai's concept of "production knowledge," which includes not only technical knowledge of an object, but also an understanding of its potential commodification and marketability, to explain how certain well-placed individuals were able to render particular natural objects marketable.

Menon uses her study of the practice of medicine to shed new light upon the question of the interaction between the English EIC as a commercial entity and its early pretensions to stateliness, which were recently highlighted by Philip Stern.[78] Her contribution focuses on the career of Scottish surgeon and botanist William Roxburgh (1751–1815) in India and his relationship with his patron, the free merchant Andrew Ross (d. 1797). Roxburgh's status within the EIC, gained on the basis of his useful knowledge, not only enabled him to amass his own private fortune, but also gave him the power to shape the emerging role of Company Naturalist. Menon's stress on the familial aspects of the patron-client relationship demonstrates that despite the emphasis on corporate sovereignty in studies such as Stern's, other forms of collaborative construction of both knowledge and power remained vital to early modern European empires. In Roxburgh's case, she shows how a particular type of knowledge-making, developed to befit the circumstances of Edinburgh—where landlords committed to the idea of "improvement" pursued the twin goals of capital accumulation and social power—was transferred to the Indian context.

Menon's chapter, like Pratik Chakrabarti's recent study,[79] also notes the military context of medical exchange. While EIC factors remained on the coast, medics travelled with armies and on diplomatic missions, allowing them to gain wider experience of the natural environment. Roxburgh's investigations into indigo for dye, the production of cotton in the Circars, and pepper plantations garnered not only financial gain but also political profit for the EIC. This was achieved in Britain partly through the celebration by the Company propagandist Alexander Dalrymple of Roxburgh's "discoveries" such as the potential of the species of indigo known as *Nerium* for dying cloth. Meanwhile Roxburgh's local influence with the *pālaiyakkārar*, or "little kings" of the Circar region through his acquisition of land aided Company objectives on the ground, enabling surveying and canal-building, even when he himself opposed the EIC's plans. Thus, as Menon shows, rather than being a peripheral to the "Company state," the medics and later naturalists in the service of the early EIC were central to its operation.

As Menon's chapter demonstrates, both the explanation of disease and the exchange of knowledge are undertaken within politicized contexts. The exchange of doctors and the display of their skills in fora such as royal courts was an important part of diplomacy in the medieval and early modern periods, and the provision of medical assistance remained a display of "soft" power into modernity, as Rashed Chowdhury's chapter also shows. Furthermore, health was not only a matter of colonial concern used as a means of reaching the areas outside of formal colonial control, but also, in some cases, a means of resisting colonial power. In Michael La Rue's chapter on the Egyptian plague of 1834–1835 the nationalist government of Muhammad Ali's standpoint reflects a commitment to independence legitimized by participation in medical modernity through enforcing unpopular quarantine measures on the Muslim population. La Rue and Jansen examine the ways in which the responses to epidemic disease illustrated societies' understandings of themselves, while La Rue and Hurgobin focus on the situation of minorities accused of spreading disease.

As La Rue shows, the plague epidemic threw light on the living conditions and occupations of sub-Saharan African slaves in Egypt, a section of the population whose lives normally passed unrecorded. Debates about the plague epidemic—notably between the French doctor, A. B. Clot, and the British politician and abolitionist, John Bowring—became inextricably intertwined with debates about the issue of slavery, as the inquiry into the causes of the transmission of plague identified the poor sanitary conditions of the slaves as a major factor. In spreading the plague from Alexandria to Cairo, however, it was another minority group—the Maltese community—that was blamed for spreading the disease. Once in Cairo, where the Jewish community came under suspicion of causing contamination, the epidemic wiped out around one-sixth of the inhabitants. Despite Bowring's convenient argument that plague was not contagious and that quarantines were unnecessary, several other European doctors in the country took the opposite view. This point of view, of course, worked to frustrate the ambitions of Muhammad Ali for a greater participation in international trade. However, it made him introduce additional public health measures in the years that followed and consolidated the place of European medical advisors, such as Clot in Egypt, with significant diplomatic consequences.

Rashed Chowdhury's chapter demonstrates how the Russian Empire in the nineteenth and twentieth centuries used medical philanthropy in Ethiopia to support its efforts to acquire a greater presence on the world stage. While the bonds between these two distant nations were based in perceived similarities between the strains of Christianity they embraced, frustrating the colonial ambitions of Italy provided a more urgent motivation for collaboration. Chowdhury's contribution highlights the hybrid nature of the Russian Red Cross, which combined official autonomy from the state with patronage from the dowager Empress Maria Feodorovna (1847–1928). The treatment of Emperor Menelik (r. 1889–1913) and his wife by the Russian medical team sent a powerful message about their recognition of the Russian medical mission as an emissary of the state. The hospitals provided by the Russians catered mainly to the lower and middle classes, but also served Ethiopian elites and church officials as well as foreigners of several different faiths. The difficulties that the Russian medical team faced when confronted with the issue of the Ethiopian rainy season reminds us that even in relatively recent history, the weather patterns of the Indian Ocean continued to suggest a seasonal rhythm to the movement of people. In his epilogue, Chowdhury demonstrates the durability of the ties of medical assistance that were established between Russia and Ethiopia, noting the revival of hospitals during the period of the Soviet Union and beyond.

Anoushka Bhattacharyya, Yoshina Hurgobin, and Cristiana Bastos and Ana Roque's chapters share a general theme of health and healing institutions as a site of interaction and struggle between various interests within the colonial state and the native population. While all the institutions they discuss were foundations of the colonial state, they were far from being homogeneous and their walls were perpetually penetrated by local beliefs and realities. In Bhattacharyya's account of "native lunacy," the boundaries between health and culture become blurred as the colonial medical officers considered entire societies to exhibit "pathological sentiments." On the other hand, she shows how "native asylums" were more culturally porous than many other colonial institutions, accepting advice from the families of patients. The permeable walls of the asylum allowed the life of the community to spill in—occasionally en masse, as with the case of the Patna asylum after the flood in 1880—and the dependence on native personnel allowed ideas about mental health to move from the grassroots up within what were normally assumed to be strongly hierarchical institutions. The hybridized institution allowed local ideas, such as the effect of spices on the mood and the assumption that the lower castes were more prone to maniacal forms of mental disorder, to seep into the colonial institution. The role of the chai wallah as a carrier of news penetrated the colonial state itself. Bhattacharyya shows how ways of working with patients developed in India spilled over into Burma through the long-distance circulation of asylum staff. At the same time, she demonstrates the contrasting lack of porosity of the asylum boundaries in the Burmese context that also arose from the transfer of personnel from India.

Kavadi's chapter focuses on the permeability and mutability of health-care institutions on a larger scale. He shows that even when experiments in public health were based on models drawn from as far off as the southern United States, their trialing within South and Southeast Asia led to the circulation of information within the region and the adaption of health-care systems to local priorities and needs. The eventual use of the Rockefeller Foundation (RF)'s rural health clinics by anticolonial nationalists in Indonesia provides an example in which the benefits of health care, distributed with one political aim, come to serve quite another. In India, the capacity building conducted by the RF enabled the development of continued local health units, although these also often evolved in different ways to those imagined by the RF.

Sugata Bose's exploration of the "circular migration" of some thirty million Indians between the 1830s and the 1930s highlighted the importance of this flow of people in the provision of capital and labour in the Indian Ocean and beyond.[80] Kavadi, Jansen, Hurgobin, and Bastos and Roque all examine the issue of migrant labor though the lens of disease and healing. Kavadi's mapping of hookworm infection in Java, Sumatra, Straits Settlement, Singapore, Siam, the Federated Malaya States, Fiji, Burma, Ceylon, and India demonstrates the correlation between the spread of disease and the movement of Indian indentured laborers. Thus, patterns of disease transmission within the Indian Ocean reflected the flow of labor that formed the connection between sites of imperial exploitation of people and natural resources. As a center for the export of indentured laborers and thus contagion, Madras eventually had to become the crux of the campaign for eradication for the RF, whose concern sprung partly from the perceived threat of hookworm spread by Indian laborers in the United States. The campaign drew in medical personnel from Australia and Mauritius, again demonstrating the connections between the networks formed by exploitation of labor, disease, and healing. However, socioeconomic conditions as well as the priorities of the colonial government made it impossible to promote sanitation measures, leading to a focus on treatment. The later concentration on drug development also led to a transition in colonial health policy beyond concerns of addressing specific health problems on the ground and toward using colonized populations as test subjects in a global experiment. This contribution takes us beyond the boundaries of the Indian Ocean, to the beginnings of a global health system, albeit one with its roots embedded in a series of very local discussions.

Hurgobin's essay makes a claim for the evolution of a specific medical ideology centering on the bodies of indentured laborers. Her focus is Mauritius: the site of migration for an estimated 455,000 Indian workers between 1834 and the end of indenture in the early twentieth century.[81] She explains the creation of epidemics with reference to the environmental factors affecting the Indian Ocean and demonstrates, through an analysis of the contemporary debates in Calcutta, how the creation of a medical ideology encompassed ideas about climate and its effects on the body. She shows how pandemics and moments of crisis could trigger oppressing laws pertaining to workers, how disease and concern over its consequences affected the transition from slavery to indentured labor, and how, on occasion, it led to the halting of the transportation of indentured workers. While philanthropic reasons were professed by the Indian industrialists who were reluctant for workers to enter into indenture in the Indian Ocean, Hurgobin convincingly argues that their economic concerns over their own supply of labor were uppermost in their minds. The interest of the Indian colonial government in the health of indentured workers was prompted partly by the concern to distinguish between this system and that of slavery. For their part, Mauritian planters were concerned to acquire workers who were bodily suited to the type of labor demanded of them. Hurgobin's chapter explores further how the condition of indentured laborers became a pawn in the struggles between different interest groups within empires.

Both Hurogobin and Jansen explore how the physical form of the island interacted with health interventions. Hurgobin shows how hospitals were used to restrict the mobility of the working classes and how islets were used as isolation wards. Her essay also takes us out onto the ocean to investigate the living quarters of workers being transported between Calcutta and Mauritius and the range of actors aboard the ship, from captain to "native doctor," who intervened in, and sometimes quarreled over, the question of the migrants' health. Jansen's investigation of the recent chikungunya epidemic in the neighboring island of Réunion uses the question of the island's physical geography in a rather different way. As Jansen notes, migration to Réunion had followed a similar pattern of European colonial settlement and an initial phase of slavery, followed by a period of indentured labor recruited primarily from India. However, the departmentalization of Réunion took it on a different path from its neighbors: its political situation acting to symbolically lift it out of its Indian Ocean context.

As Kavadi shows, imperial governments could be negligent or even obstructive when it came to making provisions for the health of their subjects. Nonetheless, if empires are to profit from the labor of those they govern, they must preserve the health of their profit-generating bodies. Thus, while conditions such as hookworm might be ignored, seriously debilitating and deadly diseases, such as malaria and the fear of contagion from sick workers, prompted the construction of institutions, such as the "Fever Hospital" in Calcutta, discussed by Hurgobin. Massive pandemics of malaria, cholera, and plague, spread further by the penetration of the railways into previous remote areas, necessitated measures to improve the health of colonial populations.

As Bose notes, the flow of migrants from India also included skilled workers who established supra-local or even global networks.[82] One such worker was Arthur Ignacio da Gama, a graduate of the Medical School of Goa, whose medical career in Sofala from 1876 is discussed by Bastos and Roque. These authors draw comparisons between da Gama and his contemporary Ezequiel da Silva, a third-generation Portuguese local teacher whose ancestors had trod the Iberian imperials circuits from Macao to Mozambique: trajectories of migration that mirrored those of earlier diaspora. During the nineteenth century, graduates of the Medical School of

Goa—an institution strictly focused on teaching Western medicine to its Indian pupils—fanned out across the Portuguese world, from the Cape Verde islands in the Atlantic to the eastern part of Timor, and were officially celebrated as agents of the empire. Bastos and Roque examine the writings of da Gama as a window into the views and experiences of one of those actively involved in this migration. These authors explore the complexities of the encounter of da Gama—who had received his medical training in an environment in which the incorporation of methods of healing drawn from local practices as pharmacopoeia was actively discouraged—with African forms of healing. What he produced is described as an ethnomedical narrative, but he, nonetheless, displayed the prejudices inherited from his European medical education. The ethnically European Ezequiel da Silva was apparently far more immersed in the local culture, as the partial result of the status of creoles in the Portuguese empire. Despite such prejudices on the part of metropolitans, men such as da Silva were useful to the colonial government by supplying helpful knowledge through his collections of local materia medica.

An interesting point raised by Bastos and Roque and by Laplante is that colonial authorities often had to try hard to de-hybridize the forms of healing that occurred in their settlements. They were concerned to mark off those forms of treatment that were considered "traditional" because they were based on "superstition" as inferior, and creating an impression of a distinctive, rational "Western" medicine at the same time. This was achieved both through institutions such as medical schools and by legal measures, such as the criminalization of certain forms of healing Laplante describes in colonial South Africa of the 1860s. This included activities such as the opening of "chemist's shops" that impinged on what was considered to be European medicine.

Kavadi and Jansen's chapters share a focus on public health. Public health campaigns, by definition, necessitate public involvement. As Arafath demonstrates, in the premodern period, incorporating messages about cleanliness into religious festivals could create public awareness about preventive measures against the spread of disease. Caste-based ideas about purity were imbued with ideas of cleanliness, and ritual hygiene and infringements were strictly punished. As he notes, the enforcement of such rules also facilitated the dominance of the affluent. Focusing on a later period, Kavadi's chapter brings to bear the interesting concept of "creating the disease in the minds of the people" in the case of hookworm infection, which also raises the larger issue of how certain conditions make the transition from being regarded as ever-present, if vexing, facts of daily life to serious threats to health.

Jansen's chapter provides a useful comparison between modern discourses of public health applied during the chikungunya epidemic in Réunion during 2005–2007 and the earlier outbreaks of malaria during the 1950s. She notes that public health discourses often reflected on the wider issues raised by the postcolonial relationship between France and Réunion. Despite the awareness during the 1950s epidemic that the poor living conditions of the majority of Réunionese was a major contributing factor, French measures to eradicate the disease focused on short-term measures such as spraying with DDT. As Jansen shows, dissatisfaction with the failure to improve living standards and eradicate disease contributed to the changing political climate in Réunion, which resulted in the separation between the aims of the *Parti Communiste Réunionnais* and its French counterpart and their campaign for independence. Conservative opponents of the idea of independence made use of the fear of the island's descent into "underdevelopment," including the threat of disease, to argue for continued attachment to France. Jansen shows how, regardless of the fact that disease will spread irrespective of political borders, the inhabitants of Réunion had, to some extent, internalized colonial ideas about disease as inherently "non-white," "tropical," or even associated with the "contaminating" effects of communism. Local interpr2etations of the disease thus mirrored the "miasmic" explanations of contagion popular in colonial discourse during the nineteenth century, as well as attempting to attach the responsibility for the disease to Mayotte and Mahoran immigrants. Such ideas, as well as the specter of terrorism that has come to haunt parts of the Indian Ocean in recent years,[83] were used to chastise the French government for its neglect of its overseas citizens.

Jansen notes the argument of Trostle[84] that disease and its interpretation should be read as an expression of society. In his highly personal exploration of discourses of healing among Zigua communities in northeastern Tanzania, Jonathan Walz shows how sickness can be an expression of the historical ills to have affected a society. Healing is thus a means of linking the past and the present and intervening in the course of the future. Walz's chapter takes us on a journey between coast and mountainous hinterland, showing how the interaction between memory and place comprises "a worldview in motion." Historical traumas, including the slave trade to the coast, colonial domination, migratory labor, and enforced "villagization" during the 1970s, are discussed using the symbolic language of serpents as spirits of ancestors and aspects of nature, including the ocean. Histories are also recreated within the process of healing, as healers travel along the historical routes of the caravan trade, putting together objects or "artifacts" gathered along the way that make reference to the slave trade, for instance, recalled by a glass bead, as well as plants with medicinal properties that are said to house ancestor spirits. The artifacts thus gathered are recombined in a medicine gourd, whose name in Swahili (*bahari*) also refers to the coast. Walz argues that this was a performative recreation that "domesticates" the past and "cools" its alien foreign influences, thereby enabling the possibility of an alternative future. Walz's experiences with the history of healing and the healing of history lead him to suggest a new type of archaeology that recognizes not only the deep-seated and long-standing links between coast and hinterland in Eastern Africa, but also local ways of interpreting these connections.

Julie Laplante's chapter also explores the interactions between localized "indigenous" beliefs about healing and their social role and the wider world, focusing on the Zulu and Xhosa healing practices collectively defined as *muti*. Like the Zigua healers encountered by Walz, the *isangoma* or Xhosa healer-divine achieves her power to cure not only through the skillful deployment of herbal medicines, but also through communication with the ancestors using activities such as drumming, dance, the interpretation of dreams, and divination. By following the preclinical trial of *Artemisia afra* conducted in Cape Town, Laplante delineates the complex relationship between local forms of healing—many of which have international roots—and biomedicine. As her chapter shows, there exists a desire among many practitioners of "traditional" medicine, as well as national governments, to demonstrate to the world that their remedies "work" through systems of validation such as biomedical testing. On the other hand, the awareness exists that the hunger on the part of pharmaceutical and agrochemical companies for local knowledge of plants and their properties often involves the unacknowledged appropriation of local knowledge.[85] Indigenous groups, while being the acknowledged guardians of some of the world's most valuable natural resources, continue to have the worst indicators for health worldwide.[86] Over the last few decades, awareness of such problems has led to the establishment of partnerships that seek to ensure that the benefits of ethnobotanical research reach the communities whose knowledge they draw on.[87] Laplante discusses one such organization, The International Center on Indigenous Phytotherapy Studies (TICIPS).

The issue of the relationship between "traditional" medicine and allopathy is further complicated in the South African context of Laplante's chapter by the interaction of indigenous medicine with Ayurvedic traditions brought by Indian settlers and Rastafarian bush doctors, who imported still more ways of engaging with plants and people in the healing process. She uses the example of the adaptation of a clinical model evolved to test Ayurvedic drugs to the evaluation of *A. afra* by TICIPS. Laplante engages the concept of an IOW at a theoretical level by employing Merleau-Ponty's "world of perception" to ask whether the idea can provide a conceptualization of the relationship between the environment and bodies that is better able to capture multiple interacting practices of healing than models such as the "One Health, One World initiative," which are based on a theoretical divide between nature and culture and often prove incapable of accommodating the less tangible aspects of healing. Laplante explores the challenges to the randomized control trial that emerged from the resistance of South Africa, Brazil, and India to global patent laws. Alternative ideas included opening up the laboratory to "let the world in" through strategies such as reversing the order of experimental testing to privilege initial clinical observations over laboratory results. Laplante also explores the complex interaction of the politics of indigeneity with the cosmopolitan actuality of "traditional" medicine in South Africa, including the importance of Indian *inyangas* and Rastafarian herbalists as well as Chinese and Malayan healers. A key commonality between the approaches of the different healers she discusses is the conceptualization of plants as sentient beings and thus as active agents in the process of healing, thus destabilizing the nature/culture divide that is central to much of Western philosophy.

All the chapters in this book explore cosmopolitan interactions with objects and practices of healing among and between communities formed through faith, language, or physical proximity, or among workers being exploited for their labor. The argument that the Indian Ocean might be regarded as a conceptual "world" rather than merely a geographical space informs our juxtaposition of these chapters that deal with very different times, places, and questions. How can we justify a claim to regard the Indian Ocean as a world in terms of its medical culture? The Oxford English Dictionary's definition of the word "world" refers to a "state or realm of existence." We might draw on Merleau-Ponty's "world of perception," mentioned by Laplante, to argue that scholars' claims for the Indian Ocean to be regarded as a world must be based on its perception as a "state or realm of existence" for those who inhabited it.

In conceptualizing the IOW, many scholars have turned to the monsoon as a uniting factor. They have emphasized the role of the ocean itself in enabling and directing encounters between the inhabitants of its various shores.[88] In an earlier work, Michael Pearson has also stressed the need to take account of the aerial as well as the aquatic aspects of the IOW as enablers of connectivity.[89] For most of the period under consideration in this book, people did not travel by airplane, but remained at the mercy of monsoon winds while crossing the seas in sailing ships. As Laplante notes, apart from being key connecting factors in the IOW, air and water are essential components and needs of the human body. To remind ourselves of the even deeper connection between the elements and the balance of fluids and vapors in the human body, perceived by many of those who thought or wrote about medicine, we might turn to Hippocrates, who lived closer to the Mediterranean Sea than the Indian Ocean, but whose ideas, nonetheless, circulated in various languages and forms across the region. His, "Airs, Waters, and Places" opens by instructing the student of medicine in any given place to begin by making environmental observations.[90] According to Hippocrates, the features of the landscape, the qualities of the waters, and the direction and strength of the winds all affect the health of the people. As each environment produced people of different dispositions, so it transmitted its qualities to the animal, vegetable, and mineral drugs that it produced. In a world of medicine based on the theory of opposites, therefore, the ideal drug to treat an imbalance of the humors might well be found an environment with opposing characteristics to those of the disease.

A world contains many worldviews, and the aim of this book is not to argue for a homogeneous medical culture that united the IOW.[91] Rather, as the importance accorded to "exotic" medicines and objects in many places across the region demonstrates, differences across healing cultures were not only recognized but also valued.[92] Currents of water and wind directed the course of expeditions around the Indian Ocean and directed the course of not only the transmission of disease but also the trade in healing objects, along with the movement of people and faiths. The balance of environmental elements was also a key factor in the conception of disease and promoted a cosmopolitan medical culture, in which the cure for a disease on one shore of the Indian Ocean was often found on another.

Notes

1. P. Gupta (2012), "Monsoon Fever," *Social Dynamics*, 38: 516–527.
2. David Arnold (1991), "The Indian Ocean as a Disease Zone, 1500–1950," *South Asia: Journal of South Asian Studies*, 14: 1–21.
3. Howard Philips (2012), "Gandhi under the Plague-Spot Light," Paper presented at the "Histories of Medicine in the Indian Ocean World" conference, Indian Ocean World Centre, Montreal, Canada, April 27, 2012.
4. Arnold, "The Indian Ocean as a Disease Zone."
5. H. Yule, A. C. Burnell, and W. Crooke (1903), *Hobson-Jobson: A Glossary of Colloquial Anglo-Indian Words and Phrases, and of Kindred Terms, Etymological, Historical, Geographical and Discursive* (London: J. Murray), p. 352, "FIRINGHEE."
6. These examples come from Leigh Chipman (2010), *The World of Pharmacy and Pharmacists in Mamlūk Cairo* (Leiden: Brill); and H. A. S. Ibn, al-Tilmīdh and Oliver Kahl (2007), *The Dispensatory of Ibn At-Tilmīḏ: Arabic Text, English Translation, Study and Glossaries* (Leiden: Brill).
7. These examples come from al-Razī's *Kitab al-Asrar*, discussed by Gail Taylor (2010), "The Kitab al-Asrar: An Alchemy Manual in Tenth-Century Persia," *Arab Studies Quarterly*, 32 (1): 6–27.
8. Fernand Braudel (1972), *The Mediterranean and the Mediterranean World in the Age of Philip II* (New York, NY: Harper & Row).

9. K. N. Chaudhuri (1985), *Trade and Civilisation in the Indian Ocean: An Economic History from the Rise of Islam to 1750* (Cambridge: Cambridge University Press); Auguste Toussaint (1961), *Histoire de l'Océan Indien* (Paris: Presses universitaires de France).
10. Chaudhuri, *Trade and Civilisation* (1985), pp. 39, 55, 66–75, 89, 185–186 and map pp. 186–187; and K N. Chaudhuri (1990), *Asia before Europe: Economy and Civilisation of the Indian Ocean from the Rise of Islam to 1750* (Cambridge: Cambridge University Press).
11. Important overviews include H. Furber, S. Arasaratnam, and K. McPherson (2004), *Maritime India* (Delhi: Oxford University Press); A. Das Gupta and M. Pearson, eds. (1987), *India and the Indian Ocean, 1500–1800* (Calcutta: Oxford University Press); M. Pearson (2005), *The World of the Indian Ocean, 1500–1800: Studies in Economic, Social and Cultural History* (Burlington: Ashgate); S. Arasaratnam (1994), *Maritime India in the Seventeenth Century* (Delhi: Oxford University Press); and S. Arasaratnam (1995), *Maritime Trade, Society and European Influence in South Asia, 1600–1800* (Aldershot: Variorum). For overviews of Indian Ocean studies, see M. Vink (2007), "Indian Ocean Studies and the 'New Thalassology,'" *Journal of Global History*, 2: 41–62; and S. Prange (2008), "Scholars and the Sea: A Historiography of the Indian Ocean," *History Compass*, 6: 1382–1393.
12. W. C. Brice (1982), *An Historical Atlas of Islam* (Leiden: E. J. Brill); I. Habib (1982), *An Atlas of the Mughal Empire* (Delhi: Oxford University Press).
13. The trans-regional aspects of Buddhist methods of healing are touched upon in a number of works. Some examples include K. G. Zysk (1991), *Asceticism and Healing in Ancient India: Medicine in the Buddhist Monastery* (New York, NY: Oxford University Press); D. R. Williams (2005), *The Other Side of Zen: A Social History of Sōtō Zen: Buddhism in Tokugawa Japan* (Princeton, NJ: Princeton University Press); P. Unschuld (2010), *Medicine in China: A History of Ideas* (Berkeley: University of California Press); J. Liyanaratne (1999), *Buddhism and Traditional Medicine in Sri Lanka* (Kelaniya: Kelaniya University).
14. A. E. Goble (2011), *Confluences of Medicine in Medieval Japan: Buddhist Healing, Chinese Knowledge, Islamic Formulas, and Wounds of War* (Honolulu: University of Hawai'i Press), especially Ch. 3.
15. A. Schottenhammer, ed. (2005), *Trade and Transfer across the East Asian Mediterranean* (Harrassiwitz Verlag: Wiesbaden).
16. See Toussaint, *Histoire de l'Océan Indien*, Ch. 5 and pp. 104–105 for the map of the Indian Ocean in the Middle Ages showing "Pays hindouises."
17. Chaudhuri, *Trade and Civilisation*, p. 100; Thomas R. Metcalf (2007), *Imperial Connections India in the Indian Ocean Arena, 1860–1920* (Berkeley: University of California Press).
18. Despite its importance, there is no overall study of the distribution of the neem tree. An indication of some of its modern uses and distribution is provided by S. Ahmed, S. Bamofleh, and M. Munshi (1989), "Cultivation

of Neem (*Azadirachta indica*, Meliaceae) in Saudi Arabia," *Economic Botany*, 43 (1): 35–38; S. A. Radwanski and G. E. Wickens (1981), "Vegetative Fallows and Potential Value of the Neem Tree (*Azadirachta indica*) in the Tropics," *Economic Botany*, 35 (4): 398–414; G. Bodeker, G. Burford, J. Chamberlain, J. R. Chamberlain, and K. K. S. Bhat (2001), "The Underexploited Medicinal Potential of *Azadirachta indica* A. Juss. (Meliaceae) and *Acacia nilotica* (L.) Willd. ex Del. (Leguminosae) in Sub-Saharan Africa: A Case for a Review of Priorities," *The International Forestry Review*, 3 (4): 285–298.

19. Sanjay Subrahmanyam (2005), *From the Tagus to the Ganges* (Oxford: Oxford University Press), pp. 45–79.
20. Engseng Ho (2006), *The Graves of Tarim: Genealogy and Mobility across the Indian Ocean* (Berkeley: University of California Press), pp. xix–xx.
21. Chaudhuri, *Trade and Civilisation*, p. 98.
22. Unschuld, *Medicine in China*; Goble, *Confluences of Medicine in Medieval Japan*.
23. M. Meyerhof (1984), *Studies in Medieval Arabic Medicine: Theory and Practice*, ed. P. Johnstone (London: Variorum Reprints); J. C. Burgel (1976), "Secular and Religious Features of Medieval Arabic Medicine," in *Asian Medical Systems: A Comparative Study*, ed. C. Leslie (Berkeley: University of California Press), pp. 44–62.
24. Ho, *The Graves of Tarim*, p. 7 and pp. 10–17 on the destruction of the tomb of "the Adeni."
25. N. Green (2012), *Making Space: Sufis and Settlers in Early Modern India* (New Delhi: Oxford University Press), Ch. 1.
26. K. Pemberton (2005), "Muslim Women Mystics and Female Authority in South Asia," in *Contesting Rituals: Islam and Practices of Identity-Making*, ed. A. Strathern and P. J. Stewart (Durham, NC: Carolina Academic Press), pp. 3–39; J. B. Flueckiger (1995), "The Vision Was of Written Words: Negotiating Authority as a Female Healer in South India," in *Syllables of Sky: Studies in South Indian Civilization*, ed. D. Shulman and V. N. Rao (Delhi: Oxford University Press).
27. For information on the modern hospital, see Iman Reza (A. S.) Network (2013), "The Holy Shrine of Emam Reza (A. S.)," http://www.imamreza.net/eng/imamreza.php?print=1077; and "Miraculous healing," http://www.imam-reza.net/eng/list.php?id=0218, both pages last accessed December 2, 2013. For a mention of the historical hospital on the site, see A. Bari and A. Hussain (2001), "Hakim 'Imād al-Dīn Mahmūd b. Fahkr al-Dīn Muhammad Shirazi," *Studies in History of Medicine and Science*, 17 (1–2): new series, 73–85.
28. E. Simpson and K. Kresse, eds. (2008), "Introduction," in *Struggling with History: Islam and Cosmopolitanism in the Western Indian Ocean* (New York, NY: Columbia University Press), pp. 1–41, quote from p. 9. Chaudhuri, *Trade and Civilisation*, p. 9, uses the same phrase: Islam created "zones of political tensions which ultimately checked its growth and destroyed the earlier sense of Arab intellectual triumph."

29. Nile Green (2008), "Moral Competition and the Thrill of the Spectacular: Recounting Catastrophe in Colonial Bombay," *South Asia Research*, 28 (3): 239–251.
30. Simpson and Kresse, "Introduction," *Struggling with History*.
31. E. Alpers (2003), "The African Diaspora in the Indian Ocean World: A Comparative Perspective," in *The African Diaspora in the Indian Ocean*, ed. Shihan de S. Jayasuriya and Richard Pankhurst (Trenton, NJ: Africa World Press), p. 30. See also A. Catlin-Jairazbhoy and E. A. Alpers, eds. (2004), *Sidis and Scholars: Essays on African Indians* (Noida, UP: Rainbow; Trenton, NJ: Red Sea Press).
32. Helen Basu (2007), "Drumming and Praying: Sidi at the Interface of Spirit Possession and Islam," in Simpson and Kresse, *Struggling with History*, pp. 291–322.
33. I. M., Lewis, Safi A. El, and Sayed H. A. Hurreiz (1991), *Women's Medicine: The Zar-Bori Cult in Africa and Beyond* (Edinburgh: Edinburgh University Press for the International African Institute).
34. H. Rangan, T. Denham, and J. Carney (2012), "Environmental History of Botanical Exchanges in the Indian Ocean World," *Environment and History*, 18 (3): 311–342.
35. M. Levey (1973), *Early Arabic Pharmacology: An Introduction Based on Ancient and Medieval Sources* (Leiden: Brill).
36. Chipman, *The World of Pharmacy and Pharmacists in Mamlūk Cairo*.
37. Hal-Tilmīdh and Kahl, *The Dispensatory of Ibn At-Tilmīd*.
38. Cyril Elgood (1951), *A Medical History of Persia, and the Eastern Caliphate: From the Earliest Times until the Year A. D. 1932* (Cambridge: Cambridge University Press).
39. Cyril Elgood (1970), *Safavid Medical Practice: Or, the Practice of Medicine, Surgery and Gynaecology in Persia between 1500 A.D. and 1750 A.D.* (London: Luzac), p. 70.
40. S. Alavi (2008), *Islam and Healing: Loss and Recovery of an Indo-Muslim Medical Tradition, 1600–1900* (New York, NY: Palgrave Macmillan); G. N. A. Attewell (2007), *Refiguring Unani Tibb: Plural Healing in Late Colonial India* (Hyderabad, India: Orient Longman); P. Mukharji (2011), "Lokman, Chholeman and Manik Pir: Multiple Frames of Institutionalising Islamic Medicine in Modern Bengal," *Social History of Medicine*, 24: 720–738.
41. F. Speziale, ed. (2012), *Hospitals in Iran and India, 1500–1950's* (Leiden and Boston, MA: Brill).
42. C. G. Uragoda (1987), *History of Medicine in Sri Lanka from the Earliest Times to 1948* (Colombo: Sri Lanka Medical Association).
43. S. M. Imamuddīn (1978), "Māristān (Hospitals) in Medieval Spain," *Islamic Studies*, 17 (1): 45–55.
44. Geert Jan van Gelder (2008), "The 'Ḥammām': A Space between Heaven and Hell," *Quaderni di Studi Arabi, Nuova Serie*, 3: 9–24.
45. F. Speziale, "Introduction," in Speziale, ed., *Hospitals in Iran and India*.

46. H. Naraindas and C. Bastos (2011), "Healing Holidays?: Itinerant Patients, Therapeutic Locals and the Quest for Health," *Anthropology and Medicine*, 18: 1–136.
47. R. H. Charlier and M.-C. P. Chaineux (2009), "The Healing Sea: A Sustainable Coastal Ocean Resource: Thalassotherapy," *Journal of Coastal Research*, 25 (4): 838–856.
48. F. Speziale, "Tradition et Réforme du dār al-šifā au Deccan," in Speziale, ed., *Hospitals in Iran and India*.
49. F. Speziale, "Introduction," in Speziale, ed., *Hospitals in Iran and India*.
50. A. Vanzan, "Hamdard, How to Share Pain in a Muslim Way," in Speziale ed., *Hospitals in Iran and India*, pp. 215–229.
51. M. Pearson (1995), "The Thin End of the Wedge: Medical Relativities as a Paradigm of Early Modern Indian-European Relations," *Modern Asian Studies*, 29 (l): 141–170; M. Pearson (1996), "First Contacts between Indian and European Medical Systems: Goa in the Sixteenth Century," in *Warm Climates and Western Medicine: The Emergence of Tropical Medicine, 1500–1900*, ed. David Arnold (Amsterdam: Editions Rodopi), pp. 20–41; M. Pearson (2001a), "Hindu Medical Practice in Sixteenth-Century Western India: Evidence from the Portuguese Records," *Portuguese Studies*, 17: 100–13; M. Pearson (2001b), "The Portuguese State and Medicine in Sixteenth Century Goa," in *The Portuguese and Socio-Cultural Changes in India, 1500–1800*, ed. K. S. Mathew, Teotónio R. de Souza, and Pius Malekandathil (Tellicherry, Kerala: Institute for Research in Social Sciences and Humanities), pp. 401–419; M. Pearson (2006), "Portuguese and Indian Medical Systems: Commonality and Superiority in the Early Modern Period," *Revista de Cultura*, 20: 116–141; M. Pearson (2011), "Medical Connections and Exchanges in the Early Modern World," in Health and Borders across Time and Cultures: China, India and the Indian Ocean Region, Special Issue of *PORTAL, Journal of Multidisciplinary International Studies* (ejournal), 8 (2): 1–15.
52. I. G. Županov (2005), *Missionary Tropics: The Catholic Frontier in India (16th–17th Centuries)* (Ann Arbor: University of Michigan Press); I. G. Županov (2006), "Goan Brahmans in the Land of Promise: Missionaries, Spies and Gentiles in the 17th–18th Century Sri Lanka," in *Portugal—Sri Lanka: 500 Years*, ed. Jorge Flores, South China and Maritime Asia Series (Roderich Ptak and Thomas O. Hölmann, eds. (Wiesbaden: Harrassowitz and the Calouste Gulbenkian Foundation), pp. 171–210.
53. C. Bastos (2010), "Medicine, Colonial Order and Local Action in Goa," in *Crossing Colonial Historiographies*, ed. Anne Digby, Waltraud Ernst, and Projit Mukharji (Newcastle: Cambridge Scholars), pp. 185–212; C. Bastos (2008), "Migrants, Settlers and Colonists: The Biopolitics of Displaced Bodies," *International Migration*, 46: 27–54; C. Bastos (2005), "Race, Medicine and the Late Portuguese Empire: The Role of Goan Colonial Physicians," *The Journal of Romance Studies*, 5 (1): 23–35.
54. H. J. Cook (2007), *Matters of Exchange: Commerce, Medicine, and Science in the Dutch Golden Age* (New Haven, CT: Yale University Press).

55. E. C. Spary (2000), *Utopia's Garden: French Natural History from Old Regime to Revolution* (Chicago, IL: University of Chicago Press); K. Raj (2007), *Relocating Modern Science: Circulation and the Construction of Knowledge in South Asia and Europe, 1650–1900* (Houndmills, Basingstoke, Hampshire: Palgrave Macmillan).
56. Just a few examples include D. Arnold (1988), "Introduction: Disease, Medicine and Empire," in *Imperial Medicine and Indigenous Societies*, ed. D. Arnold (Manchester: Manchester University Press); M. Harrison (1994), *Public Health in British India: Anglo-Indian Preventive Medicine, 1859–1914* (Cambridge: Cambridge University Press); M. Harrison (1999), *Climates & Constitutions: Health, Race, Environment and British Imperialism in India, 1600–1850* (New Delhi, Oxford University Press); J. Iliffe (1998), *East African Doctors: A History of the Modern Profession* (Cambridge, Cambridge University Press).
57. David Hardiman, ed. (2006), "Introduction," *Healing Bodies, Saving Souls: Medical Missions in Asia and Africa* (Amsterdam: Rodopi), p. 1.
58. D. Arnold, ed. (1996), *Warm Climates and Western Medicine: The Emergence of Tropical Medicine, 1500–1900* (Amsterdam: Editions Rodopi), p. 11.
59. Braudel c.f. Chaudhuri, *Trade and Civilisation*, p. 221.
60. Ernst Waltraud and Projit B. Mukharji (2009), "From History of Colonial Medicine to Plural Medicine in a Global Perspective," *NTM International Journal of History and Ethics of Natural Sciences, Technology and Medicine*, 17: 447–458.
61. See also Muzaffer Alam (2004), "Sharia, Akhlaaq & Governance," in *The Languages of Political Islam: India, 1200–1800*, ed. Muzaffar Alam (Chicago, IL: University of Chicago Press).
62. Attewell, *Reconfiguring Unani* Tibb, Ch. 6.
63. Sherman Cochran (2006), *Chinese Medicine Men: Consumer Culture in China and Southeast Asia* (Cambridge, MA: Harvard University Press).
64. Deepak Kumar (1995), *Science and the Raj, 1857–1905* (Delhi: Oxford University Press).
65. E. De Michelis (2004), *A History of Modern Yoga, Patanjali and Western Esotericism* (New York, NY: Continuum International).
66. D. Wujastyk and F. Smith (2008), *Modern and Global Ayurveda: Pluralism and Paradigms* (Albany: State University of New York Press).
67. Sienna R. Craig (2012), *Healing Elements: Efficacy and the Social Ecologies of Tibetan Medicine* (Berkeley: University of California Press).
68. Francis Zimmermann (1992), "Gentle Purge: The Flower Power of Ayurveda," in *Paths to Asian Medical Knowledge*, ed. Charles M. Leslie and Allan Young (Berkeley: University of California Press).
69. Paul U. Unschuld, "Epistemological Issues and Changing Legitimation: Traditional Chinese Medicine in the Twentieth Century," in Leslie and Young, *Paths to Asian Medical Knowledge*; Attewell, *Refiguring Unani Tibb*.
70. Karen Fjelstad and Nguyen T. Hien (2011), *Spirits without Borders: Vietnamese Spirit Mediums in a Transnational Age* (New York, NY: Palgrave Macmillan).

71. See L. Monnais and D. Wright, eds. (2015), *The Transnational Migration of Physicians in the Twentieth Century* (Toronto: University of Toronto Press), forthcoming; and the related research project, "Medical Diasporas," http://www.medicaldiasporas.org, last accessed February 16, 2015.
72. O. Dyer (1993), "Gynaecologist Struck off Over Female Circumcision," *British Medical Journal*, 307 (6917), 1441–1442; June Kelly (February 4, 2015), "Doctor Found Not Guilty of Performing FGM," BBC News, http://www.bbc.co.uk/news/uk-england-31138218, last accessed February 16, 2015.
73. See Naraindas and Bastos (2011) and several papers in the special issue of *Anthropology and Medicine*, 18, now also published as an edited book: H. Naraindas and C. Bastos (2014), *Healing Holidays: Itinerate Patients, Theraputic Locales and the Quest for Health* (Oxford: Routledge).
74. Lyn Schumaker, Diana Jeater, and Tracy Luedke (2007), "Histories of Healing: Past and Present Medical Practices in Africa and the Diaspora," *Journal of Southern African Studies*, 33: 707–714, and other papers in this special issue.
75. Thomas Kuhn (1970), *The Structure of Scientific Revolutions* (Chicago, IL: University of Chicago Press).
76. Raj, *Relocating Modern Science*.
77. Cook, *Matters of Exchange*.
78. Philip J. Stern (2011), *The Company-State: Corporate Sovereignty & the Early Modern Foundations of the British Empire in India* (Oxford and New York, NY: Oxford University Press).
79. Pratik Chakrabarti (2010), *Materials and Medicine: Trade, Conquest, and Therapeutics in the Eighteenth Century* (Manchester: Manchester University Press).
80. Sugata Bose (2006), *A Hundred Horizons: The Indian Ocean in the Age of Global Empire* (Cambridge, MA: Harvard University Press).
81. Ibid., p. 77.
82. Ibid., p. 73.
83. On the Indian Ocean as a key strategic area in global politics, see Robert Kaplan (2010), *Monsoon: The Indian Ocean and the Future of American Power* (New York, NY: Random House).
84. J. A. Trostle (2005), *Epidemiology and Culture* (New York, NY: Cambridge University Press).
85. John Merson (2000), "Bio-Prospecting or Bio-Piracy: Intellectual Property Rights and Biodiversity in a Colonial and Postcolonial Context," *Osiris*, 2nd series, 15: 282–296.
86. Carolyn Stephens et al. (2006), "Disappearing, Displaced, and Undervalued: A Call to Action for Indigenous Health Worldwide," *The Lancet*, 367 (9527): 2019–2028.
87. Merson, "Bio-Prospecting or Bio-Piracy."
88. Gupta, "Monsoon Fever."
89. Michael Pearson (2010), "The Idea of the Indian Ocean," in *Eyes across the Water: Navigating the Indian Ocean*, ed. Pamila Gupta, Isabel Hofmeyr, and Michael Pearson (Pretoria: UNISA Press).

90. (G. E. R. Hippocrates), John Chadwick Lloyd, and W. N. Mann, eds. (1983), *Hippocratic Writings* (Harmondsworth: Penguin).
91. See Sanjay Subrahmanyam, "Persianisation and 'Mercantilism' in Bay of Bengal History, 1400–1700," in *From the Tagus to the Ganges*, p. 52 for a critique of the simplistic views of Indian Ocean trade post-750 as dominated by a homogeneous form of Islam.
92. For cosmopolitanism as the celebration of difference, see Kwame Anthony Appiah (2006), *Cosmopolitanism: Ethics in a World of Strangers* (New York, NY: W. W. Norton).

1

An Insight into al-Razi's Extraordinary Theoretical and Practical Contributions for Developing Arthrology

Mahmud Angrini

Introduction

Abu Bakr Muhammad ibn Zakariyā Rāzī (250 AH/864 AD–311 AH/923 AD), known as al-Razi, or Rhazes in the West, and often referred to as "the Galen of the Arabs,"[1] has been considered one of the most important and famous doctors throughout the ages since his lifetime. He has been referred to as "the Physician par excellence of the Muslims"[2] and the "Learned Master (*Allamah*) of the Sciences of the Ancients."[3] He was a physician, philosopher, and a chemist. In medicine, his contributions were so significant that they can only be compared to those of Ibn Sina (Avicenna).[4] He was taught by a number of teachers from Khurasan: al-Nishaburi,[5] Abu Zayd al-Balkhi,[6] and Ali ibn Rabban al-Tabari.[7] He made a thorough study of medical practice in the hospitals of Rayy and Baghdad, finally achieving the rank of director of the largest hospital (*al-Bimaristan al-Aodedi*) in Baghdad and also becoming a court physician. al-Razi was renowned across the Indian Ocean world. He debated with many outstanding personalities of his time. He discussed metaphysics and the problem of time with Abu al-Qasim al-Ka'bi al-Balkhi;[8] the question of the preexistence of matter with Ahmad ibn al-Hasan al-Masma'i;[9] the validity of medicine with Abu al-Abbas al-Nashi al-Akbar;[10] the problem of pleasure with Abu al-Hasan Shahid ibn al-Husayn al-Balkhi;[11] that of bitter tastes with Ahmad ibn Muhammad Abu Tayyib al-Sarakhsi;[12] and the question of the inanimate with Ahmad ibn Kayyal.[13] The esteem in which al-Razi was held is indicated by the *Fihrist* ("bibliography") of his writings compiled by Abi Rayhan al-Biruni, another great learned man of the fifth/eleventh century.[14] al-Biruni undertook the work despite his strong disagreement with al-Razi on a number of philosophical and religious issues. al-Razi is well known for his scientific methodology and for his efforts to practice medicine on the basis of clinical research. Although al-Hawi is considered his most famous collection of writings, he excelled in the formulation of many other compositions such as *Kitab al-Mansuri fi al-tibb* ("The Medical Book of Mansur"), *al-Fakher* ("The Magnificent Book"), *al-Shokouk Ala Jalinus* ("The Doubts upon Galen"), and *Risaleh fi al-Hasba and al-Jadari* ("A Treatise of Measles and Smallpox") in addition to his treatise titled *fi Awjaa' al-Mafasel* ("On Joint Pains").

Of more than 200 compositions, mostly in Arabic, attributed to the famous Muslim Persian physician, al-Razi, the Arabic treatise *fi Awjaa' al-Mafasel* ("On Joint Pains") occupies a very special position, because it is considered one of the first written works that specialized in arthrology, and may be the oldest that survives up until now. Despite the importance of this treatise, it has not yet received enough attention from historians of medicine. The reasons for this neglect include the absence of an edited copy of this Arabic manuscript, and the location of all the manuscript copies in either Tehran or London: out of reach for many researchers of Arabic medicine. Nonetheless, many physicians in Indian Ocean terrains, including al-Samarqandi and Shams al-Din al-Laboudi, followed al-Razi's lead in composing independent works in the field of arthrology and quoted his thoughts.

Initially, a distinction should be made between two of al-Razi's works: "On Joint Pains," the focus of this paper, and another manuscript titled *Maqāla fī al-naqras* ("On Gout"), which is a completely different manuscript, edited by Zidan in 2003.[15] Many of the historians of Islamic medicine of the modern era have confused one manuscript with the other. For example, in his chapter devoted to al-Razi, Hamarneh writes: "As a conclusion of this chapter, we mention another small treatise about gout and joints' pains".[16] Through his description of the manuscript copies and the number of its chapters, it becomes evident that it is actually the treatise "On Gout." Zuhair Humaidan has made a similar mistake, because he mentions, among the works of al-Razi, "The Book of Gout and Joints' Pains," listing three copies of the manuscript. Of these, the first two are manuscripts of the treatise "On Gout" and the last is a copy of the treatise "On Joint Pains."[17] Dr. Ahmet Agirakca similarly mentions among the books of al-Razi "The Book of the Illnesses of Joints and Gout," and then when he lists the manuscript copies, he mentions the copy in the Municipal Library in Alexandria, which belongs to the treatise "On Gout" and the copy of Malek library in Tehran, which belongs to the treatise "On Joints Pains."[18]

Attribution of the Manuscript *fi Awjaa' al-Mafasel* ("On Joint Pains") to al-Razi

In his famous book, *Kitab Ikhbar al-'ulama' bi-akhbar al-hukama* ("The History of Learned Men"), usually referred to simply as *Ta'rikh al-hukama*, al-Qifti (lived c. 1172–1248 CE) mentioned "The Book of Joint Pains" in his list of al-Razi's works.[19]

Ibn Abi Usaybi'ah (1203–1270), who wrote a very important book named *Uyūn ul-Anbā' fī Ṭabaqāt ul-Aṭibbā* ("Lives of the Physicians"), mentioned "The Book of Illnesses of Joints, Gout and Sciatica" among the literature of al-Razi. He thought that it consisted of 22 chapters.[20]

al-Biruni confirmed the attribution of this treatise to al-Razi in his treatise *Risalat Abi Rayhan fi fihrist kutub al-Razi* ("A treatise of the works cited of Muhammad ibn Zakariya al-Razi" or "al-Biruni's Comprehensive Bibliography of the Works of al-Razi"), which was edited and published by B. Kraus in 1936, based on the only copy that is preserved in the library of Leiden.[21]

Finally, at the beginning of the manuscript "On Joint Pains," we find the following statement, which is present in all three copies and was written by the copiers themselves: "A treatise of al-Razi on joint pains and their treatment,"[22] which confirms the attribution of the manuscript to al-Razi.

Manuscript Copies in International Libraries

According to Sezgin,[23] Brokelmen,[24] and Ramadan Shashen et al.,[25] there are only three known copies of this manuscript in the libraries of the world:

1. A copy in the Malek Library in Tehran, Iran: It is catalogued as 4573/8i.[26] It was copied by Mansour bin Wli-allah in 1086 AH (1675–1676 AD). This copy consists of 13 pages, each page containing about 43 lines of text, each line including an average of 20 words.[27]
2. A second copy in the library of Malek Library in Tehran, Iran: It is catalogued as 4442/i.[28] It was copied by Ibn Haj Maasoum in 1243 AH (1827–1828 AD). This copy consists of 60 pages, each page containing about 15 lines of text, each line consisting of an average of 12 words.[29]
3. The third copy is the thirtieth chapter of a large manuscript titled "The important chapters in the nation's medicine" and was collected by a physician named Ibn Sharabion Ibrahim al-Mutatabbeb. It is preserved in the library of Cambridge University, Cambridge, the United Kingdom. It is catalogued as 3516.[30] There is no information regarding the date of the copy and the copier. The whole manuscript consists of 235 pages, and the thirtieth chapter consists of 33 pages. Every page contains about 25 lines, each line consisting of an average of 14 words (Figures 1.1 and 1.2).

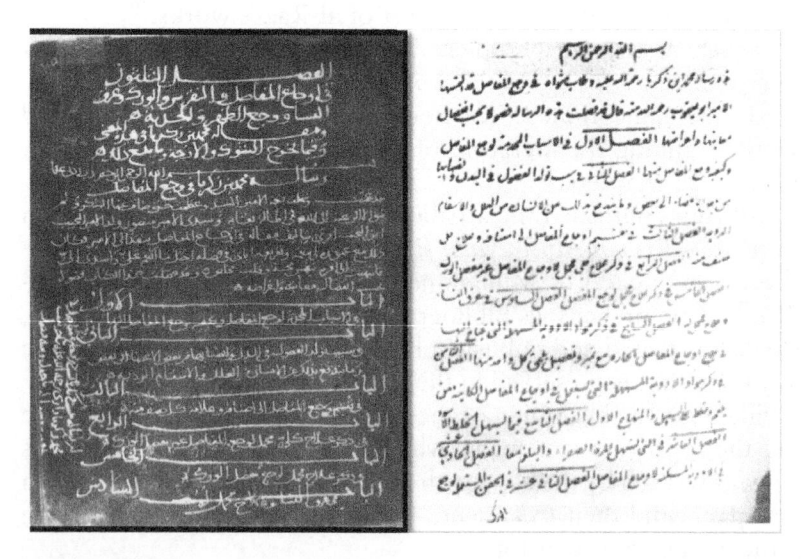

Figure 1.1 The first page of the manuscript. Left: the copy in the library of Cambridge University (MS 3516); right: the copy in the Malek Library in Tehran, Iran (MS 4442/i).

Figure 1.2 The first page of the second copy in the library of Malek Library in Tehran, Iran (MS 4573/8i).

The differences between the three copies in terms of the number of lines and pages is actually mainly due to the different orthography used in each of them. The copyists made several errors in handwriting. In rare instances we find some additions in one copy that are not present in the others. One such example is the addition of a whole case study in the Malek Library copy in Tehran (MS 4442/i).

The Significance of the Manuscript for the History of Medicine

Upon the examination of the English translation of the Hippocratic works,[31] a treatise called "On the Articulations" can easily be noticed. However, the scientific content of this piece is mainly concerned with orthopedic surgery: it discusses fractures, dislocations, spinal injuries, and other conditions requiring surgical intervention; thus, it does not discuss joint pains or what is now called rheumatology. Galen, in turn, did not compose a specialized book or treatise on arthralgia. This conclusion was reached after the examination of the lists of Galen's works by Ibn Abi Usaibia[32] and Jamaluddin al-al-Qifti.[33] The same conclusion was reached from an examination of the letter of Hunain bin Ishaq to Yahya bin Ali regarding the works of Galen that had been translated into Arabic.[34]

One of the books that al-Razi depended on as a resource for his treatise was Rufus's book "On Joint Pains," as it was called in al-Razi's *al-Hawi*.[35] "A treatise of the diseases of joints" is mentioned by Ibn Abi Usaibia.[36] This book no longer exists according to Sezgin.[37] Thabit ibn Qurra (d. 288 AH) wrote a book titled "A Treatise on Joint Pain" according to Ibn Abi Usaibia.[38] This book was undoubtedly a very important resource for al-Razi's treatise.[39] However, according to Sezgin, no intact copies of this manuscript are in existence.[40]

This makes it probable that al-Razi's treatise is the oldest Arabic medical book in rheumatology that still exists. Many physicians followed al-Razi's lead and wrote specialized treatises on joint pains. One of them was Najib el-Din Samarqandi (d. 619), who wrote "A treatise on treating joint pain."[41] Shams el-Din Muhammad Ben al-Laboudi, the physician of Damascus (570–621 AH), also wrote a specialized treatise on joint pain.[42]

al-Razi's Goal in Writing This Treatise

At the beginning of the treatise, al-Razi states that it was composed at the request of Prince Mansur: "My master, the Prince Mansur of noble descent, had told me to compile a treatise on joint pains for his sake, and that was done with his grace and virtues."[43] Later in the text, it becomes clear that al-Razi's underlying motive for composing the treatise was to respond to doctors who argued that the cause of joint pains was that the joints in question are weak in nature. In contrast, al-Razi himself believed that the pain was caused by the effusion of humors to the joints.[44]

al-Razi took a long time to complete his treatise and he drew on several resources to do so. It seems that al-Razi composed his treatise "On Joint Pains" after he wrote his book "Introduction to Medicine Industry or Isagogic" because al-Razi said at the conclusion of his treatise:

> I wanted to begin this treatise by a chapter dedicated for the definition of medical terms used in this treatise...but because I explained these terms in the book that I had made as an introduction to the medical industry, I thought that composing of such chapter in this treatise is a surplus.[45]

It is probable that al-Razi had composed this treatise before he wrote *al-Mansuri*. In the dedication of the treatise to Prince Abu Saleh Mansur, he attempted to ingratiate himself in overblown fashion, declaring: "The grace of the Prince has pervaded and overflown to the degree that the thanksgiving was unable to compensate, leaving only the desire to God in prolonging his survival."[46] al-Razi followed a similar approach in his other works, before being criticized by his contemporaries for his attempts to get closer to the governors. This criticism is mentioned in al-Razi's work *al-Syrat al-Falsafiah* ("The Philosophical Approach"), where al-Razi wrote: "Some people who have perspective, distinction, and achievements blamed and criticized us when they saw us engaging with people and making a living, and they claimed that we strayed from the line of the great philosophers, particularly that of our imam, Socrates, because it is well known that he did not visit Kings, and he used to underestimate them if they paid him a visit."[47]

After writing this passage, al-Razi abandoned his overblown prefaces and began to introduce his books with only simple words such as those that begin the *al-Mansuri*: "For the sake of the honored prince, Mansur ibn Ishak, I gathered statements, compilations and anecdotes of the medical industry in my book, and I try to seek abbreviation and brevity."[48]

The Contents and Message of the Manuscript "On Joint Pains"

al-Razi's treatise "On Joint Pains" is composed of an introduction and 22 chapters. By examining the contents of the chapters in the treatise, it can be concluded that more than half of the treatise was

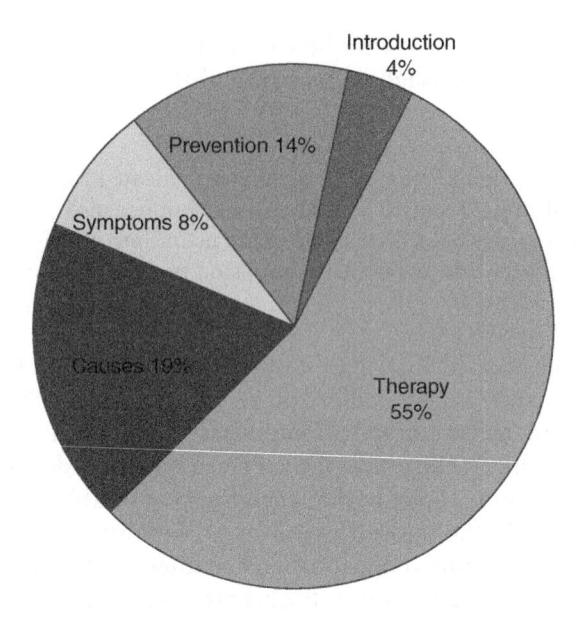

Figure 1.3 The approximate percentages for the contents of the treatise.

devoted to the various methods of treatments he used to cure arthralgia. Figure 1.3 clarifies how al-Razi divided up his treatise and how many lines he dedicated to every section.

Causes of Joint Pain According to al-Razi

al-Razi hypothesized an integrated and cohesive theory based upon humoral theory to explain arthralgia, to interpret its symptoms, and to indicate suitable treatments. His hypothesis was opposed, according to his own statement, to the common belief among his contemporary physicians who believed that the cause of arthralgia is merely the inherent weakness of the joints in question.

Humoral theory proposed that our bodies are composed of four humors: black bile, blood, phlegm, and yellow bile. These four humors were considered to be the basic units and fundamental building blocks of all natural organisms. In good health it was believed that our humors were in harmonious balance throughout the body. Illness was thought to be the manifestation of that balance being disturbed. The concept of four humors may have origins in Ancient Egyptian medicine or in Mesopotamia, although it was not systemized until ancient Greek thinkers[49] directly linked it with the popular theory of the four elements—earth, fire, water, and air—around 400 BC. The development of humoral theory is associated originally with Hippocrates (ca. 460–370 BCE). In the second century CE, Galen elaborated on this theory, which was further developed by Arabic writers beginning in the ninth century and by European writers beginning in the eleventh century. The four humors and the importance of keeping them in balance in order to maintain good health was the most influential of the theories that were passed down from the Greeks and Romans; it remained a dominant factor in medicine until the nineteenth century.[50]

In the first chapter of his treatise, al-Razi argues that the primary cause of joint pain is the accumulation of the excretion of digestion in veins. Other less important reasons include the accumulation of the excretion of digestion in the liver, followed by other excretions. The powerful internal organs push those excretions forward to vulnerable organs, most notably the joints, through special outlets, which in turn cause arthralgia due to either an increase in the amount of excretions or the change in their nature, or both.[51] al-Razi considered joints to be sites designed for accepting the excretions from powerful organs, because as machines for flexion and extension, they are required to contain blank spaces.[52]

In the second chapter of his treatise, al-Razi posits that understanding the physiology and the anatomy of the abdomen and the digestive organs is essential to understand the mechanism of the formation of humors. This, he thought, would help the reader "to well understand our following arguments and to make them more convincing."[53] In this chapter, we outline many interesting points.

First, al-Razi describes what is now known as the "hepatic portal system":

> All of these branches connect to each other and the smaller branches join the bigger until they form the great tract which is called *Bab al-Kabed* [the portal vein]. All those branches absorb nutrients from the intestine and push them from the smaller to the larger branches, until they lead all the absorbed nutrients to the Portal vein, and this happens in a way that resembles the reunion of small streams to form rivers which in turn join together to form seas or swamps. Then, this portal vein divides and ramifies to very many branches in

all the regions of the liver, until they become very thin. Then, they contribute to the branches which will form the origin of the Great Vein [he may mean the inferior vena cava] which will sprout out from the hepatic tuber which we are going to mention.[54]

Second, the treatise contains a description of blood capillaries. al-Razi writes:

> Many branches ramified off it.[55] Those branches irrigate all the inferior organs, and they are well known and able to be recognized during the autopsy, until they become very small divisions and they end up as veins known as capillaries because they are too thin and they are mixed with the meat giving it its red color.[56]

Although it is now widely accepted that Marcello Malpighi discovered blood capillaries when he used a microscope in 1661 to examine the brain and major organs to demonstrate their finer anatomical features, many former physicians observed and described capillaries: Leonardo da Vinci may be one of them.[57] al-Razi's observations predate those of both Malpighi and da Vinci by centuries. Nonetheless, credit for the complete understanding of the function of the capillaries might still be accorded to Marcello Malpighi and many scientists who followed him.

As for al-Razi's predecessors, Galen believed that the blood moved from veins to arteries through small invisible bores located in the interventricular septum of the heart,[58] so he did not recognize the presence of capillaries. al-Tabari in his book *Fardos al-Hekma* (The Paradise of Wisdom) did not pay attention to the capillaries in the section he dedicated to the circulation of blood.[59] Therefore, al-Razi's observations seem likely to have been based on his own experience.

Third, when he recommended the venesection of the sciatic vein (the small saphenous vein) as a therapy for sciatica pain, al-Razi referred to the possibility of venesecting its branches on the lateral dorsal side of the foot in case of the invisibility of the sciatic vein itself. He wrote:

> The cure for this type is to venesect the vein known as the sciatica from where it appears on the lateral side of the leg, and in case it did not appear, we can venesect some of its branches on the lateral half of dorsal foot, I mean the region of the ring toe [fourth toe] and the

pinky toe [fifth toe] because the branches in this area diverge from this vein.[60]

This description is correct, in that the small saphenous vein arises from the lateral part of the dorsal venous arch of the foot.[61] Galen described the sciatic vein (the small saphenous vein) as follows: "The second vein is the outer one; it descended upon the smaller bone of the leg until it reaches the joint of malleolus."[62] This demonstrates that Galen believed that the small saphenous vein ends at the lateral malleolus, and does not connect with any branches on the dorsal feet. Therefore, once again, it is quite probable that al-Razi developed this idea from his own experience and not from Galen's works on anatomy.

The Clinical Description of Arthralgia in the Manuscript

In the third part of his treatise, al-Razi defined the clinical bases on which he relied for his division of joint pains into types. He justified his interest in the classification of joint pains by the following argument:

> However, we should emphasize the terminology of this field: investigate it, explain and clarify it, because the whole matter depends on it, and because most often the error happens in recognizing the disease not in its treatment in case it was well determined.[63]

al-Razi believed in the possibility of recognizing the effused humor in joints by means of observing a patient. As he wrote: "The skin color of the joint, its palpation, the type of pain, the type of treatment which can relieve or irritate it, his former medicines, his age, his country and his present and former illnesses."[64] These passages from al-Razi's writings evince the diagnostic strategies that he used to narrow the diagnostic possibilities and determine the exact type of arthralgia present.

al-Razi considered that the pain in the joints is caused by an effusion of one of the humors: blood, bilious humor, phlegm, or a mix of them. He considered that the effusion of melancholic humor (black bile) usually caused sciatica, and rarely other joint pains.[65] The theory of al-Razi is summarized in Figure 1.4.

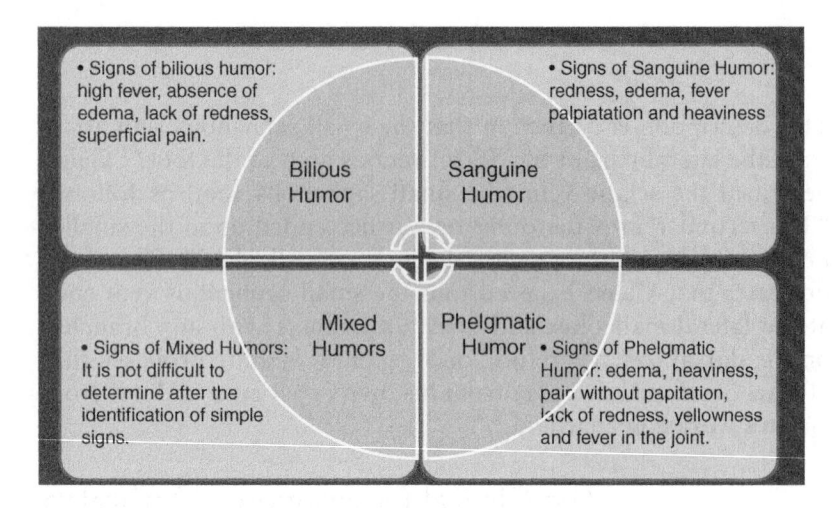

• Signs of bilious humor: high fever, absence of edema, lack of redness, superficial pain. Bilious Humor	• Signs of Sanguine Humor: redness, edema, fever palpiatation and heaviness Sanguine Humor
Mixed Humors • Signs of Mixed Humors: It is not difficult to determine after the identification of simple signs.	Phelgmatic Humor • Signs of Phelgmatic Humor: edema, heaviness, pain without papitation, lack of redness, yellowness and fever in the joint.

Figure 1.4 The theory that al-Razi depended on to explain arthralgia.[66]

al-Razi was interested in following the concept of the clinical case study in his treatise. He introduced into his essay discussions of six clinical cases, the development of which he carefully tracked from one day to another, giving careful and detailed explanations, descriptions, and interpretations of each case.

Sciatica Pain According to al-Razi

al-Razi defined sciatica by saying: "This is a severe pain [that] travels along the length of [a] patient's leg, from the hip to the toes like a stretched tendon. Sometimes, it doesn't reach the foot and it interrupts at the malleolus or at the knee."[67] It is clear that this clinical description is absolutely true and in accordance with the standards of modern medicine. But al-Razi thought that the reason for this pain was a fullness in the veins of the leg (not the nerves) either because of the accumulation of pure blood or because of a melancholic humor at the lower section of the body. This would be accompanied by numbness and coldness in the lower extremity.

al-Razi thought that the treatment of sciatica should depend upon "the venesection of vein known as the sciatic vein from where it appears from the lateral side of the leg or from some of its branches on the dorsal foot."[68] In addition, he recommended medicines for eliminating the melancholic humor if it was the cause of the fullness of veins, as well as painkillers and even the use of cauterization. al-Razi recommended his patients to avoid long walks and horse riding.[69]

Treatment of Arthralgia in the Manuscript

al-Razi follows a set of principles in describing treatments for each type of joint pain. Some of those principles are general and were developed by some predecessors, such as Hippocrates, Galen, Ali ibn Rubban al-Tabari, and Sabur Ibn Sahl, but others were actually developed by al-Razi himself. Prior to al-Razi, Sabur Ibn Sahl (d. 869) wrote the oldest handwritten witness to Arabic pharmacy known so far, called *al-Aqrabadhin* ("Dispensatory"),[70] which was divided into 22 chapters. He discussed the general principles of prescribing drugs, but his prescription was not well organized as that of al-Razi. Those principles formed a standard that was followed by later scholars such as Ibn Sina (Avicenna) and Ibn al-Tilmidh. al-Razi's advice regarding the suitable treatment of joint pains differed according to the type of the responsible humor. In other words, he tried to correlate treatment with the pathology of the pain. For example, while he recommended cooling dressings[71] and venesecting the veins from the opposite side of the injured joint to treat pains resulting from bloody effusions,[72] he recommended laxatives and applying moisturizing dressings upon the injured joint if the humor was bilious.[73]

al-Razi advised that the use of painkillers should be restricted to severe cases only, and he only mentioned the medicines that he had tested and tried himself.[74] al-Razi dedicated two separate chapters of the treatise to the treatment of pain in the hip joint. Chapter V focuses on drugs and Chapter XII focuses on the injections used in the treatment. He justifies his allocation of so much space to the problem on the following grounds: "The hip joint is a deep joint, covered with a large amount of meat, which makes it difficult to infer the caused humor from the color and the palpation of the joint."[75] He also notes that, unlike the other joints, the occurrence of hip dislocations and lameness, as a complication, is common. In contrast, it is rare to experience joint contracture and ossification.[76]

al-Razi also stated that cool and hot dressings are of little benefit in hip joints because their effects cannot reach the inner depth of the hip joint.[77] In Chapter XII, al-Razi lists seven recipes for suppositories used for the treatment of pain in the hip joint.[78] al-Razi allocated five chapters (VII–XI) to the laxatives and sedatives that are needed for the treatment of joint pain, in accordance with the type of humor that he considered had poured into the joint. He followed a very systematic approach in describing these medicines.

al-Razi describes unfamiliar and new drugs in detail, while he gives abbreviated references to well-known drugs.[79] He starts with simple drugs, then moves on to the compound medicines. He first mentions mildly effective drugs, then moves on to the more effective drugs. This method is based on his conviction in the necessity of starting treatment with mildly effective drugs because they have fewer side effects. He writes: "Let us recall now the laxatives of phlegm, starting from the mild drugs to the more powerful in a similar way to what we did before."[80]

al-Razi described the ingredients and the method of preparation for each drug in detail. Sometimes, he mentioned two or more prescriptions that share the same ingredients but differ in the amounts of ingredients and the method of preparation. His interest in the methods of preparation of medicines is evident in these sections of the text. Sometimes, he states the best time to take a particular medicine in addition to the conditions that can help to enhance its impact. For example, he advised that a patient should take a laxative for melancholic bile at bedtime,[81] and to avoid eating for seven hours after taking another drug.[82]

al-Razi left the most powerful drug that can cure all types of arthritis until the last lines of the chapter, because of its side effects. He writes:

> Although al-Soranjan [Colchicum autumnal][83] is advantageous to all types of joints' pain, it affects the libido and causes the contraction of joints, so it is highly recommended to those who are obliged to consume a large amount of it, to take something to protect their stomachs.[84]

al-Razi realized that some compound medications can be prepared by treating them with water or alcohol, while others cannot be dissolved in water, and should therefore be treated with vinegar. This demonstrates that al-Razi's proficiency in chemistry aided him in his pharmacology. In cases in which he had been unable to test the drug, al-Razi tended to admit that fact. For example, in Chapter X, he wrote: "Several ancients mentioned that the plant named *Regel al-Gorab* [*Carum Roxburghianum*][85] is more useful than al-Soranjan to cure this malady and in spite of that it does not harm the stomach, but I have not tested it yet."[86] al-Razi considered the side effects of a drug when he prescribed it, especially when the regimen was designed to continue for a long time. For example, when he wanted to prescribe a drug known to cause strong dehydration, he advised the patient to swallow it once every two days in all months of the year except June, July, and the first half of August[87].

al-Razi devoted some chapters to specific treatments, for example, he allocated Chapters XIV and XV to discussing warm and cold ointments, and Chapters XVII and XVIII to describe the cupping therapy used to relieve the pains of the hip joint. He described in detail the device (the sucking cup) that he used to treat the pains of this special joint. Plus, he elaborately described treatment by cauterization. Although the final Chapter XXII is one of the shortest, it is an important chapter because it demonstrates how al-Razi disregarded the use of talismans in treatment, although they were very familiar in his era as well as having been adopted by many ancient physicians. He said:

> Our tendency toward treating patients with these things is very weak, but because many famous ancient physicians mentioned them and because we wanted our essay to include all types of treatments that are mentioned and described for the treatment of this pain, we dedicated this chapter to mention only those which have received more unanimity and have the more reliable sources.[88]

After this introduction, al-Razi mentions only four talismans. Furthermore, he ascribes every talisman to the ancient physician who advised the use of it, in contrast to his method in the rest of the essay where he rarely specifies the source of his medications. The effect is to emphasize that these talismans are not prescriptions he would have recommended personally.

Prevention of Joint Pain in the Manuscript

Like Hippocrates, al-Razi believed that it is better to prevent than to cure. He simply tried to apply the famous principles of the maintenance of health to the field of arthrology. According to al-Razi, the principle for the prevention of joint pains is to prevent the formation of bad humors. In case of their formation, the aim is to get rid of them quickly and without any delay to avoid their accumulation and damage in and to the main organs, which will later be forced to push them into the weakest organs such as the joints[89].

For al-Razi, prevention is based on the following points:

- *Food and drink*: al-Razi focused on the need to avoid foods that generate the humor that usually causes the type of joint pain that the patient complains of most. For example, he advised patients to prevent joint pain resulting from the bloody humor by avoiding foods such as meats, alcohols, sweets, and highly nutritious foods.[90] He also emphasized the need not to allow patients to reach a state of ultimate satiety; should that occur, patients should be encouraged to vomit. It is evident that al-Razi suggested that a good diet and weight control can help to prevent joint pain, and this is partially correct.
- *Sleep, wakefulness, comfort, fatigue, and sports*: al-Razi asserted the importance of experiencing them in moderate amounts.
- *Sexual activities*: al-Razi believed that sexual intercourse is harmful for all types of joint pain, especially pains in the lower limbs and hips.
- *Bathing*: He advised moderate bathing, especially in the times of expected pains.

al-Razi recommends monitoring the patient's pulse, mien, and his skin color on a regular basis to detect any irregularities in the body and to correct them before they lead to joint pains. Some of his recommendations were directed toward special types of arthritis in particular. For example, he advised using hot baths for arthralgia resulting from the phlegmatic humor.[91]

Unlike the other types of joint pains, al-Razi advised using emetics more than laxatives for the prevention of pain in the hip joint and sciatica. He recommended avoiding sexual activities, walking or riding for a long time, and leaning on the afflicted hip joint when standing or sleeping.[92]

To prevent sciatica, al-Razi advised patients to avoid walking, lifting heavy weights, and all work where they would be obliged to lean on their legs intensely.[93] This clinical observation corresponds entirely with what we know nowadays. al-Razi also advised patients to experience a lot of sweating in the steam bath (*hammam*), which was probably a contributing factor in relieving the accompanying muscle spasm.

Arthrology in al-Razi's Other Works

Arthrology in al-Hawi

It is now recognized that *al-Hawi* is not a formal medical encyclopedia, but a posthumous compilation of al-Razi's working notebooks, which included knowledge gathered from other books as well as original observations on diseases and therapies, based on his own clinical experience.

al-Razi devoted more than 180 pages of Part XI of the *al-Hawi* to discussing joint pains, sciatica, gout, aching hips, knees, and back, and other conditions.[94] However, the pages he dedicated to arthralgia in *al-Hawi* lack order and do not follow a logical sequence of ideas. He starts by discussing treatment of joint pains,[95] then he moves quickly to differentiate between gout, joint pains, and sciatica,[96] explains sciatica,[97] heads toward treating the pain of the hip joint,[98] mentions a dressing for gout and sciatica,[99] and so on. al-Razi continues this unorganized style throughout the entire section. al-Razi did not dedicate this section to joint pains alone, and it lacks the physiological and anatomical explanation of arthralgia, which can be found in his dedicated treatise about joint pains. Furthermore, it also lacks the special theory that he proposed to define the causes of arthralgia. al-Razi focused on the therapy of joint pains, while he almost ignored the question of prevention. The various treatments and recipes that he mentions composed more than 90 percent of the material in the section.

al-Razi mentions many of his informants (31 former physicians), some of whose opinions he quoted in this long section, in addition to his personal opinions. Sometimes, he even mentions the exact book he quoted from, also giving the name of the writer adapted according to the fashion of his times. The following authors produced works that al-Razi relied on: Hippocrates, Galen, Rufus, Othorseves, Paul, Alexander, Aristotle, Shimon the monk, Oribasius,

Vlegros, Tiadhuq, Qusta ibn Luqa, Georges, Arcaganish, Hermes, Yuhanna Ibn Masawayh, Ibn Serapion (Johannes Serapion), Ahron the priest, al-Tabari, al-Kendi, Hunain ibn Ishaq, Thabit ibn Qurra, Masarjawaih, Masih of Damasian, Abu Greg, Isaac Ben Hunain, al-Saher, Bukhtishu, Abdus, al-Jewish, Sabor, and "unknown." It is worth mentioning that he relied more on Hippocrates and Galen (especially the book translated as *Kitāb al-Mayāmir* or "The Book of Homilies"), Rufus, and Yuhanna ibn Masawayh than the rest. He also lists a number of previous works on gout, but only mentions two former books specialized in joint pains: the book of Rufus and the book of Thabit ibn Qurra.

As for his use of earlier sources, al-Razi criticizes Hippocrates as a defender of the idea that the reason for arthralgia is that the joints are inherently weak. He quotes what Hippocrates wrote in the second essay of his *Prognostics*:[100] "Diseases often happen in joints because of their capacity and the continuity of their movement, which make them one of the weaker positions."[101] al-Razi then outlines his own contrary opinion.

One very important point in the part regarding joints pains in *al-Hawi* is related to the anatomical positions of the veins of the lower extremity. When al-Razi discussed the preferred site for performing venesection to treat the pains of the lower extremity's joints, he opposed Galen's opinion. Galen thought that only one vein enters the lower extremity, then it separates to the saphenous (great saphenous vein) and the sciatic vein (small saphenous vein) at the back of the knee (Figure 1.5). al-Razi believed that there are two veins in the thigh, not one as Galen had proposed. According to recent anatomical knowledge, the great saphenous vein joins the common femoral vein in the region of the femoral triangle at the saphenous-femoral junction,[102] while the small saphenous veins usually drain into the popliteal vein, at or above the level of the knee joint.[103] Consequently, al-Razi was more accurate than Galen in his anatomical description. al-Razi said in *al-Hawi*: "Galen mentioned that unlike the upper limb, only one vein come to the lower limb, and it divides at the knee, but the experiment witnesses what we say."[104] To verify what al-Razi quoted of Galen, I reviewed an Arabic translation of Galen's anatomical treatises "for beginners," and it seems that al-Razi was exact in what he quoted.[105]

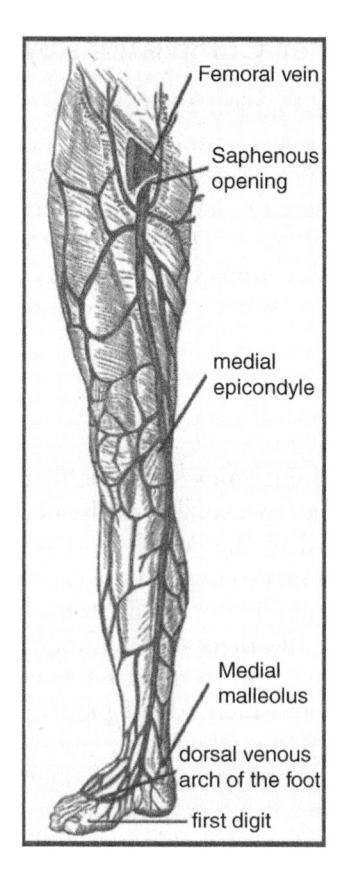

Figure 1.5 Great saphenous vein.
Source: *Gray's Anatomy of the Human Body*, 20th US Edition, Lea & Febiger, Philadelphia, 1918. Reproduced from http://en.wikipedia.org/wiki/Great_saphenous_vein.

Arthrology in *the Kitab al-Mansuri fi al-Tibb*

al-Razi allocated only a small portion (three pages) of his Arabic work *Kitab al-Mansuri fi al-tibb* ("The Medical Book of Mansur") to the discussion of gout and sciatica and hip and joint pains.[106] His description was brief and concise. It is obvious that al-Razi did not expatiate on the subject because he had already composed his treatise titled "On Joint Pains" and presented it to Prince Mansur. He confirms this point of view when he says at the end of this section: "Discussing this subject is outside the scope of this book."[107]

Arthrology in Other Compilations by al-Razi

As was the case in the *Kitab al-Mansuri fi al-Tibb* ("The Medical Book of Mansur"), al-Razi continued to neglect arthralgia in many of his subsequent Arabic works such as *al-Tibb al-mulūkī* ("The Royal Medicine"), where he devoted only three pages to the subject,[108] and *Taqāsīm al-'ilal* ("The Divisions of Diseases"), where he dedicated two small chapters to the subject, including nothing new.[109] In *Maa alfark* ("What's the Difference?") he omitted the subject entirely.[110]

Conclusion

al-Razi's treatise "On Joint Pains" is probably the oldest remaining work, at least in Arabic, specializing in the field of arthralgia. The methodology employed in this article was somehow classical, in terms of al-Razi's attempts to explain the causes of diseases, drawing on his deep understanding of the anatomy of joints and related organs, depending on the theory of humors. Having established this basis, he then moved on to discuss the symptoms of the diseases, the treatments, and methods of prevention. al-Razi formulated a coherent theory to explain joint pains, and he almost tried to explain all his thoughts within the framework of this theory. He was courageous in introducing his ideas, although they contradicted what many former doctors, including authorities such as Hippocrates, had already declared. The treatise is notable for al-Razi's description of the intestinal blood circulation and blood capillaries and the origin of the small saphenous vein (sciatic vein) in addition to his opposition to Galen regarding the tract of veins in the lower extremities. Despite his respect for Galen and Hippocrates, al-Razi dared to criticize some of their opinions. This unprecedented act was a controversial issue among both his contemporaries, such as al-Biruni (Alberonius), and his followers, such as Ibn Sina (Avicenna). They could not accept his audacity in criticizing those who were considered "The Gods of Medicine."

Except for this article and *al-Hawi*, most other works of al-Razi that I reviewed lacked any important scientific material related to joint pains. This might be explained by al-Razi's general tendency not to discuss at length the diseases to which he had formerly allocated specialized treatises; however, this idea requires further investigation. As noted above, it is now understood that *al-Hawi* represents only a notebook that al-Razi used to record his own conclusions and many excerpts of other physicians' works that attracted his attention. In was intended as a preliminary step toward using those notes to compose other treatises and books. Further research might reveal more about the method of composition of al-Razi's other works.

al-Razi did not separate joint pains from gout in any of his other works except for these two specialized treatises about joint pains and gout. al-Razi did not generally mention the sources that he relied upon in his treatise and in the parts of the other books that he dedicated for joint pains except those of *al-Hawi*. We cannot say, by any means, that the scientific materials of the treatise are contained in *al-Hawi*, because *al-Hawi* mostly concentrates on treatments and lacks the integrated theory and the adequate explanation for the causes and the symptoms of diseases, as well as the discussion of means of prevention that are present in "On Joint Pains." In addition, the treatise contains important additions to *al-Hawi*, and vice versa. While there are some discrepancies in their medical information, the two texts also contain a lot of similar material.

al-Razi enriched his treatise with some clinical cases. This reflects his desire to pay attention to the importance of following up patients and recording the changes in their health status due to the disease and the treatments as well. al-Razi also followed a scientific method in the prescription of pharmacological treatments, and provided an important clinical description of sciatica (herniated nucleus pulpous disease) and the ways to prevent it. His emphasis on careful observation and experimentation is evident in his dissatisfaction with the use of talismans in the treatment of joint pains.

There is no doubt that the *al-Hawi* remains the most important existing work by al-Razi. However, the specialized treatises that al-Razi wrote, including his treatise on joint pains and their treatment, embrace several scientific additions. Examining the treatises clarifies the methodology of al-Razi in the composition of medical works and his comprehensive view of diseases. Therefore, it is no longer acceptable to consider that studying *al-Hawi* is sufficient to achieve a complete knowledge of what al-Razi wrote concerning any particular illness.

Notes

1. Aḥmad ibn al-Qāsim Ibn Abī Uṣaybiʻah, Nazar Riḍā, ed. (1965), *Uyūn al-anbā' fī ṭabaqāt al-aṭibbā (The Sources and Information of the Classes of Physicians)* (Beirut: Dār-al-Ḥayāh Library), p. 415.
2. Djamāl al-Dīn Abu l-Ḥasan al-Qifṭī J. Lippert, ed. (1903), *Ta'rīkh al-ḥukamā': Wa-huwa Mukhtaṣar al-Zwzanī al-musammā bi l-Muntakhabāt al-multaqaṭāt min kitāb Ikhbār al-ʻulamā' bi-akhbār al-ḥukamā* (History of Learned Men) (Leipzig), p. 271.
3. Jamāl al-Dīn Abī al-Maḥāsin Yūsuf ibn Taghrībirdī Fahīm Muḥammad Shaltūt, ed. (1929–1956 [1348–1369 AH]), *al-Nujūm al-zāhirah fī mulūk Miṣr wa-al-Qāhirah*, 12 vols. (Cairo: Dār al-Kutub al-Miṣrīyah), vol. 3, p. 209.
4. Sharif Kaf al-Ghazal (2007), *The Valuable Contributions of al-Razi (Rhazes) in the History of Pharmacy* (Manchester: Foundation for Science Technology and Civilization), p. 2, http://www.muslimheritage.com/uploads/The_Valuable_Contributions_of_al-Razi_in_the_History_of_Pharmacy.pdf, last accessed February 18, 2015).
5. Nāṣir-i Khusraw Muhammad Badhl al- Rahmam, ed. (1941), *Zād-i Musātirīn* (Berlin: Kaviani), p. 98.
6. Muḥammad ibn Isḥāq Ibn al-Nadīm, Gustav Flügel, ed. (1964), *Al-fihrist* (Beirut: Maktabat Khayyat), p. 299.
7. al-Qifti, *Ta'rīkh al-ḥukamā'*, p. 231.
8. Aḥmad ibn Yaḥyá Ibn al-Ibn al-Murtada, Susana Dival-Vilzar, ed. (1961 [1380 AH]), *Kitāb Ṭabaqāt Al-Muʻtazilah* (Beirut: al-Matbaʻah al-Kāthūlīkīyah [in kommission bei Frantz Steiner Verlag]), p. 88.
9. ʻAlī Ibn al-Ḥusayn al-Masʻūdī (1965), *Kitāb at-tanbīh wa-ăl-išrāf* (Beirut: Maktabat Khayyat), p. 342.
10. al-Murtada, *Kitāb Ṭabaqāt Al-Muʻtazilah*, p. 93.
11. Yāqūt ibn awi, Allāh al-Ḥamawī, Ferdinand Wüstenfeld, ed. (1866), *Muʻjam al-buldān (Geographical Dictionary)*, 6 vols. (Leipzig: F. A. Brockhaus), vol. 2, p. 168.
12. Yāqūt ibn ʻAbd Allāh al-Ḥamawī D. S. Margoliouth, ed. (1924), *Irshád al-aríb ilá maʻrifat al-adíb* (Dictionary of Learned Men) (Cairo), vol. 1, p. 158.
13. Mutahhar b. Tahir al-Maqdisi, Clément Huart, ed. (1899–1919), *al-Bad' wa-'t-ta'rīḫ*, 6 vols. (Paris), vol. 5, p. 124.
14. Muḥammad ibn Aḥmad al-Bīrūnī, Paul Kraus, ed. (1936), *Risālah lil-Bīrūnī fī fihrist kutub Muḥammad ibn Zakarīyā' al-Rāzī* (Paris: Matbaʻat al-Qalam).
15. Abu Bakr Mohammed bin Zakaria al-Razi, Yūsuf Zaydān, ed. (2003), *A Treatise on Gout = Traîté sur la goutte = Abhandlung zur gicht = Maqāla fī al-naqras* (Alexandria, Egypt: Library of Alexandria).

16. Sami K. Hamarneh (1986), *History of the Heritage of Medical Sciences of the Arabs and Muslims* (Jordan: Publications of Yarmouk University), vol. 1, p. 214.
17. Zuhair Ḥumaidān (1996), *A'lām al-haḍāra al-'arabīya al-islāmīya fi 'l-'ulūm al-asāsīya wa't-taṭbīqīya*, 6 vols *(Figures in Arab and Islamic Civilization in the Basic and Applied Sciences)*, 6 vols. (Dimašq [Damascus]: Manšūrāt Wizārat at-Taqāfa (Ministry of Culture), vol. 2, p. 367.
18. Agırakça Ahmet (2010), *İslâm tıp tarihi (History of Islamic Medicine)* (Istanbul, Turkey: Akdem Yayinlari), p. 176.
19. al-Qifti, *Ta'rīkh al-ḥukamā'*, p. 180.
20. Aḥmad ibn al-Qāsim Ibn Abī Uṣaybi'ah, Muḥammad Bāsil 'Uyūn al-Sūd, ed. (1998),'*Uyūn al-anbā' fī ṭabaqāt al-aṭibbā (Lives of the Physicians)* (Beirut: Dār al-Kutub al-'Ilmīyah), p. 375.
21. Abū Rayḥān Muḥammad ibn Aḥmad Abi Rayhan al-Bīrūnī, Paul Kraus, ed. (1936), *Risālah lil-Bīrūnī fī fihrist kutub Muḥammad ibn Zakarīyā' al-Rāzi* ("al-Biruni's Comprehensive Bibliography of the Works of al-Razi"), Islamic Medicine Publications, Part 25, Muhammad ibn Zakariya al-al-Razi Texts and Studies 2, reprint and collected by Fuad Sezgin, Institute of the Arab-Islamic History of Science, Frankfurt, Germany, p. 2.
22. Abu Bakr Mohammed bin Zakaria al-Razi, manuscript of a treatise titled "On Joint Pains and their Treatment," Malik Library, Tehran, Iran, MS 4573/i8, paper 1B
23. Fuat Sezgin (1970), *Geschichte Des Arabischen Schrifttums*, Band III: Medizin, Pharmazie, Zoologie, Tierheilkunde—bis ca. 430 H. (Leiden: E. J. Brill), p. 284.
24. Karl Brokelman, Abdul Halim al-Najjar, trans. (1977), *Tarikh al-Adab al-'Arabi* (History of Arabic Literature), 5th ed., 6 vols. (Cairo: Dar al-Maaref).
25. Ramazan Şeşen, Cemil Akpınar, Ekmeleddin İhsanoğlu, and Cevad İzgi (1984), *Catalogue of Islamic Medical Manuscripts in the Libraries of Turkey* (Istanbul: Research Centre for Islamic History, Art and Culture).
26. Sezgin, *Geschichte Des Arabischen Schrifttums*, III, p. 284
27. al-Razi, On Joint Pains, Malik Library, Tehran, Iran, MS 4573/i8, paper 1B.
28. Sezgin, *Geschichte Des Arabischen Schrifttums*, III, p. 284.
29. al-Razi, On Joint Pains, Malik Library, Tehran, Iran, MS 4442/i, paper 1B.
30. Index of Micro-filmed Manuscripts (Supplement) in the Library of the Institute of the Arab Scientific Heritage, prepared by Mohamed Ezzat Omar, p. 97.
31. Hippocrates, Francis Adams, trans. (1952), *Hippocratic Writings* (Chicago, IL: Encyclopedia Britannica).
32. Ibn Abī Uṣaybi'ah, al-Sūd, ed.,'*Uyūn al-anbā' fī ṭabaqāt al-aṭibbā*, pp. 78–113.
33. al-Qifti (1908), *Ekhbar Alolama' be Akhbar Alhukama* (Informing Scholars of the Stories of Philosophers) (Egypt: Dar al-Saadah), pp. 85–92.
34. Ḥunain ibn Isḥāq, "Ḥunain ibn Isḥāq message to Yahya bin Ali regarding the books of Galen translated into Arabic," Ḥunain ibn Isḥāq (d. 260/873): Texts

and Studies, Vol. 15 of *Manshūrāt Ma'had Tārīkh al-'Ulūm al-'Arabīyah wa-al-Islāmīyah: Ṭibb al-Islāmī*; repr. in Fuat Sezgin (1996), *Galen in the Arabic Tradition, Texts and Studies*, Vol. 1 (Frankfurt am Main: Institute for the History of Arab-Islamic Science), pp. 206–258.

35. Al-Razi, Abu Bakr Mohammad bin Zakarya (1962), *Kitāb al-Ḥāwī fī ṭ-ṭibb* (Dā'irat al-ma'ārif al-'uṭmānīya: Ḥaidarābād ad-Dakkān [Press of Ottoman Educational Ministery in Hyderabad, India]), Part 11, 1st ed., p. 216.
36. Ibn Abī Uṣaybi'ah, *Uyūn al-anbā' fī ṭabaqāt al-aṭibbā*, p. 36
37. Sezgin, *Geschichte Des Arabischen Schrifttums*, III, p. 66.
38. Ibn Abī Uṣaybi'ah, *Uyūn al-anbā' fī ṭabaqāt al-aṭibbā*, p. 257.
39. al-Razi, *al-Ḥāwī fī ṭ-ṭibb*, pp. 235, 245.
40. Sezgin, *Geschichte Des Arabischen Schrifttums*, III, p. 377.
41. Ḥumaidān, *A'lām al-ḥaḍāra al-'arabīya al-islāmīya*, IV, p. 269.
42. Ibid., p. 183.
43. al-Razi, "On Joint Pains," Malik Library, Tehran, Iran, MS 4573/i8, paper 1B.
44. Ibid., paper 12B.
45. Ibid., paper 13A.
46. Ibid., paper 1B.
47. al-Razi, A treatise "On Gout," p. 11.
48. al-Razi Abu Bakr Mohammad bin Zakarya, Hazem al-Bakri, ed. (1987), *al-Manṣūrī fī al-ṭibb (al-Mansur on Medicine)* (al-Kuwayt: Ma'had al-Makhṭūṭāt al-'Arabīyah, al-Munaẓẓamah al-'Arabīyah lil-Tarbiyah wa-al-Thaqāfah wa-al-'Ulūm [Kuwait: Publications of the Institute of Arabic Manuscripts]), 1st ed., p. 11.
49. Karl Sudhoff (1926), *Essays in the History of Medicine* (New York, NY: Medical Life Press), p. 67.
50. Kate Kelly (2009), *The History of Medicine: The Middle Ages: 500–1450* (New York, NY: Facts On File), p. 2.
51. al-Razi, "On Joint Pains," Malik Library, MS 4573/i8, paper 1B.
52. Ibid.
53. al-Razi, "On Joint Pains," Malik Library, MS 4573/i8, paper 2A.
54. al-Razi, "On Joint Pains," Malik Library, MS 4573/i8, paper 2B.
55. He refers to the arterial branches of the hepatic artery.
56. al-Razi, "On Joint Pains," Malik Library, MS 4573/i8, paper 2B.
57. J. M. Pearce (2007), "Malpighi and the Discovery of Capillaries," *European Neurology*, 58(4): 253–255.
58. Charlotte Hwa and William C. Aird (2007), "The History of the Capillary Wall: Doctors, Dscoveries, and Debates," *American Journal of Physiology—Heart and Circulatory Physiology*, 293 (2): H2667–H2679.
59. 'Alī ibn Sahl Rabbān al-Ṭabarī, M. Z. Siddiqi, ed. (1928), *Firdaus al-ḥikma fi 'ṭ-ṭibb (Paradise of Wisdom in Medicine)*, Afthab edition (Berlin, Germany: Buch- u. Kunstdruckerei "Sonne"), pp. 334–336.
60. al-Razi, "On Joint Pains," Malik Library, Tehran, Iran, MS 4573/i8, paper 6B.
61. Richard Snell (1992), *Clinical Anatomy, Upper and Lower Extremities*, translated into Arabic by Aktham Khatib and Rafael Amid (Damascus, Syria: Dar al-Shady), 4th ed., p. 216.

62. Galen, Muhammed Nazem Mahrouseh, ed. (2009), *Jawamea' Kitab Jalinus fi al-Tashrih lemotaalmeen* (the Arabic translation of the compilation of Galen's books on the Anatomy for Beginners), Unpublished Master's thesis, Aleppo University, p. 136
63. al-Razi, "On Joints Pains," Malik Library, Tehran, Iran, MS 4573/i8, paper 4A.
64. Ibid.
65. Ibid., paper 6 B
66. al-Razi, "On Joint Pains," Malik Library, MS 4573/i8, paper 4B.
67. Ibid.
68. Ibid.
69. Ibid.
70. Sābūr Ibn Sahl, Oliver Kahl, ed. (2003), *The Small Dispensatory* (Leiden: Brill).
71. al-Razi, "On Joint Pains", Malik Library MS (4573/i8), paper 5 B
72. Ibid., paper 4B.
73. Ibid., paper 5 B
74. Ibid., paper 9 B
75. Ibid., paper 6 A
76. Ibid., paper 6 B
77. Ibid.
78. Ibid., paper 10 A
79. Ibid., paper 7 A
80. Ibid., paper 9 A
81. Ibid., paper 9 B
82. Ibid., paper 12 B
83. Colchicum autumnal contains colchicine, a useful drug with a narrow therapeutic index and many side effects. No antidote for its toxicity is presently known. Joanne Barnes, Linda A. Anderson, and J. David Phillipson (2007), *Herbal Medicines*, 3rd ed. (London and Grayslake, IL: Pharmaceutical Press), p. 29.
84. al-Razi, "On Joint Pains," Malik Library, MS 4573/i8, paper 9B.
85. Aḥmad 'Īsā, *Dictionnaire Des Noms Des Plantes En Latin, Français, Anglais Et Arabe =: Mu'ǧam Asmā' An-Nabāt* (Le Caire: Impr. Nationale), 1930, p. 40.
86. al-Razi, "On Joint Pains," Malik Library MS 4573/i8, paper 8B.
87. Ibid., paper 12B.
88. Ibid., paper 13B.
89. Ibid., paper 10A.
90. Ibid.
91. Ibid., paper 12A.
92. Ibid., paper 12B
93. Ibid.
94. al-Razi, *al-Hāwī fī ṭ-ṭibb*, Part 11, pp. 97–280.
95. Ibid., p. 97.
96. Ibid., p. 98.

97. Ibid., p. 99.
98. Ibid., p. 100.
99. Ibid., p. 102.
100. The Prognostics of Hippocrates was translated into Arabic by Hunain ibn Ishak, who named it "Takdemet al-Maarefah."
101. al-Razi, *al-Hāwī fī t-tibb*, Part 11, p. 107.
102. Wikipedia, "Great saphenous vein," http://web.archive.org/web/20150219114717/http://en.wikipedia.org/wiki/Great_saphenous_vein, accessed and archived February 19, 2015.
103. Wikipedia, "Small saphenous vein," http://web.archive.org/web/20150219115722/http://en.wikipedia.org/wiki/Small_saphenous_vein, accessed and archived February 19, 2015.
104. al-Razi, *al-Hāwī fī t-tibb*, Part 11, p. 101.
105. Galen, Muhammed Nazem Mahrouseh, ed. *Jawamea' Kitab Jalinus fi al-Tashrih lemotaalmeen* (The compilation of Galen's books on the Anatomy for Beginners), p. 136.
106. Al-Razi, Al-Mansouri, pp. 453–455.
107. Ibid., p. 455.
108. Abu Bakr Mohammed bin Zakaria al-Razi, Muḥammad Yāsir Zakkūr, ed. (2009), *al-Ṭibb al-mulūkī* (The Royal Medicine) (Beirut, Lebanon: Dar al-Manhaj), pp. 248–251.
109. Abu Bakr Mohammed bin Zakaria al-Razi, Sobhi Mahmoud Hemmami, ed. (1992), *Taqāsīm al-'ilal'* (The Divisions of Diseases) (Syria: Publications of the University of Aleppo), pp. 482–491.
110. al-Razi Abu Bakr Mohammed bin Zakaria, Salman Katayeh, ed. (1978), *Mā Alfark?* ("What's the Difference?") (Syria: Publications of the University of Aleppo), 1st ed.

2

Exchanges and Transformations in Gendered Medicine on the Maritime Silk Road: Evidence from the Thirteenth-Century *Java Sea Wreck*

Amanda Respess and Lisa C. Niziolek

The *Java Sea Wreck* collection housed at The Field Museum of Natural History in Chicago contains the excavated cargo of a wrecked thirteenth-century trading vessel found off the Indonesian coast (Figure 2.1). Excavated in 1996, the artifacts from the wreck represent a cross-section of goods traded across the span of the Indian Ocean world (IOW) and beyond, linking communities in East Africa, the Middle East, and South, Southeast, and East Asia. The *Java Sea Wreck* artifacts shed light on the cultural exchanges along water routes that have come to be known as the Maritime Silk Road or Porcelain Road[1] owing to coastal China's dominance in the production of porcelain in the period (Figure 2.2). Setting sail during the interstitial transition between the Southern Song (1127–1279 CE) and Yuan Dynasties (1271–1368 CE) and carrying a load heavy with ceramics, the ill-fated *Java Sea Wreck* vessel likely embarked from the Chinese port city of Quanzhou in Fujian province during the late thirteenth century.[2]

This chapter provides a brief inventory of the medical artifacts from the *Java Sea Wreck* within the context of emerging women's reproductive medicine in Song and Yuan period China. Additionally, the material culture of nonmedical objects from the wreck, bearing

Figure 2.1 Location of the *Java Sea Wreck* (adapted from Mathers and Flecker, 1997: 2); map redrawn by David Quednau.

design motifs and symbolism representative of the medical philosophies exchanged along the Porcelain Road, is discussed. An analysis of these objects in the context of thirteenth-century medical texts concerning fertility and women's health exposes an intercultural dialogue between Chinese and Near and Middle Eastern medicine at the dawn of the specialization of *fuke*, women's medicine, in China.

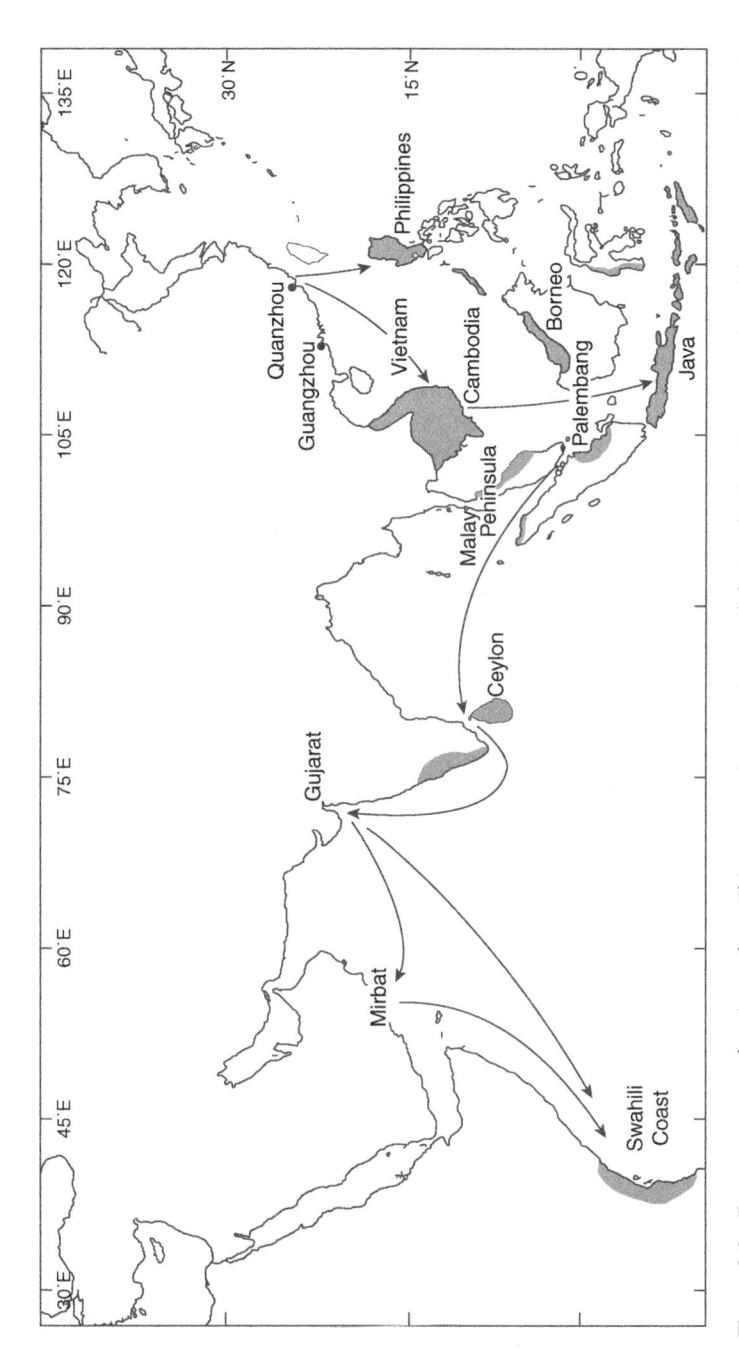

Figure 2.2 Export routes and regions where Chinese ceramics were imported during the Song Dynasty (adapted from Brown, 1997: 103); map redrawn by David Quednau.

Chinese Medical Thought along the Maritime Silk Road during the Song Dynasty

The maritime exchange that propelled the voyage of the *Java Sea Wreck* vessel coincided with a massive reformation in the annals of Chinese medicine, which Furth has described as being "reshaped" completely during the 300-year span of the Song Dynasty (960–1279 CE).[3] Medical diagnostics and treatment methods rooted in Yin-Yang theory were refined in the period into increasingly precise, standardized specializations of official medicine. The Song Dynasty witnessed the birth of fuke, women's medicine—literally *wives'* medicine—which Furth describes as the "first time it was possible to identify literate experts who were specialists...and identify a body of doctrine that related disorders deemed specific to women to the broader themes of medical discourse."[4] The broadest themes of classical Chinese medical discourse and primary organizational strategies for diagnosis and treatment underwent standardization in this period. Zhang Zhongjing's third-century medical treatise, now known as the *Jin Gui Yao Lue Fang Lun* (*Synopsis of Formulas of the Golden Chamber*), was compiled and codified by members of the Northern Song Imperial Academy and Bureau of Censoring, Proofreading, and Publishing Medical Books. Along with Zhang's *Shang Han Lun* (*Treatise on Febrile Diseases*) it has shaped Chinese medical theory and practice from the Song Dynasty onward.[5]

Yin and Yang

In the second volume of his masterwork, *Science and Civilization in China*, Needham draws a connection between a detached, Taoist perceptual habit and observation-based scientific thinking.[6] Sivin recaps this point, writing that Needham credits a Taoist "disinterested empirical observation of Nature" with the emergence of scientific theories and technologies in China.[7] Taoist observations of nature played a major role in Song Dynasty textual constructions of health and fertility operating within the natural cosmos, with the Yin-Yang imagery of the Taiji symbol depicting "correspondence between the trigrams, the cosmos and the human body."[8] Derived from observations of the landscape, the character for yin,

阴 (陰), depicts the shaded side of a hill, and yang, 阳 (陽), the side illuminated by the sun.

Yin and yang were promoted in Song Dynasty medicine as contrasting but complementary organizing principles used to categorize bodily and cosmological relationships. Central to fuke, yin signified all that is female, shaded, quiescent, substantive, and earthly, while yang signified all that is male, bright, active, ethereal, and heavenly.[9] The diagnostic and etiological framework of the Song Dynasty codification of the *Shang Han Lun* relied on the five principles of Yin-Yang philosophy, which are outlined as follows by Wiseman and Ellis:[10]

1. All phenomena have yin and yang aspects.
2. All yin and yang aspects can be further divided into yin and yang.
3. Yin and yang give rise to each other.
4. Yin and yang control each other.
5. Yin and yang transform into each other.

Crucial to Song constructions of gender and fertility, Yin-Yang medical symbolism is found in multiple aspects of the *Java Sea Wreck* collection. We will begin our inventory of the relevant cargo here, returning to the intercultural context of Porcelain Road medical philosophies as we progress.

Cargo of the *Java Sea Wreck*

Discovered in the 1980s by fishermen and subjected to several years of looting, the *Java Sea Wreck* was formally excavated in 1996.[11,12] After completing the archaeological recovery of the remnants of the vessel and its contents, Pacific Sea Resources donated half of the material found to The Field Museum, delivering the other half to the Indonesian government. The *Java Sea Wreck* and its associated cargo testify to the flourishing interregional trade that took place in the South China Sea region during the Song Dynasty. The city of Quanzhou in Fujian province—the likely port of departure of the *Java Sea Wreck* vessel—became China's largest port after the relocation of the Chinese court from northern to southern China. While China exported massive quantities of silk, ceramics, and metals to Southeast Asia and the IOW, it imported pearls, spices, medicinals, aromatics, and other delicacies and materials from across the region.

The boat itself was likely a lash-lugged vessel of Indonesian origin and is estimated to have been approximately 35 meters long and 10 meters wide.[13] Its cargo was largely made up of Chinese products—primarily iron and ceramics. Although the majority of ceramics are high-fired wares from China, there are a significant number of pieces of non-Chinese origin, in particular Indian-style *kendis* and *kundikas*. The ship's wreckage also contained other manufactured and natural commodities, including bronze figurines, finials, trays, and gongs; copper and tin ingots; elephant tusks; aromatic resin; and a single canarium nut and a degraded nipa palm fruit.[14] Based on ceramic styles, radiocarbon dating of a piece of resin found at the wreck site, and cyclical date stamps on two storage jars, the ship is believed to have sailed during the second half of the thirteenth century, a transitional period between the Song and the Yuan Dynasties.[15]

Java Sea Wreck Medicinals

Though the majority of organic goods once aboard the *Java Sea Wreck* vessel have long since disintegrated under seawater, many of the remaining artifacts have significance for the context of thirteenth-century maritime medicinal trade. The 16 pieces of unworked elephant tusk in the cargo are of unknown origin, but would have likely been exports from Vietnam, Thailand, or Sumatra, or else reexported from African or Indian sources (Figures 2.3 and 2.4) via the Arab ivory trade.[16]

During the Song and Yuan Dynasties, ivory was imported into China for artistic and decorative uses as well as for medicinal purposes. Laufer documents the presence of powdered ivory as a constituent ingredient in the Chinese medicinal *long gu* (dragon's bone), a traditional remedy comprised of crushed animal bones. Laufer dates the appearance of long gu in Chinese materia medica to the Tang Dynasty (618–907 CE), when China's maritime trade to the south and west dramatically expanded access to foreign goods.[17,18] Ivory powder was used in the treatment of "fevers, hemorrhages, and fluxes," and would "act on the liver" as a cordial

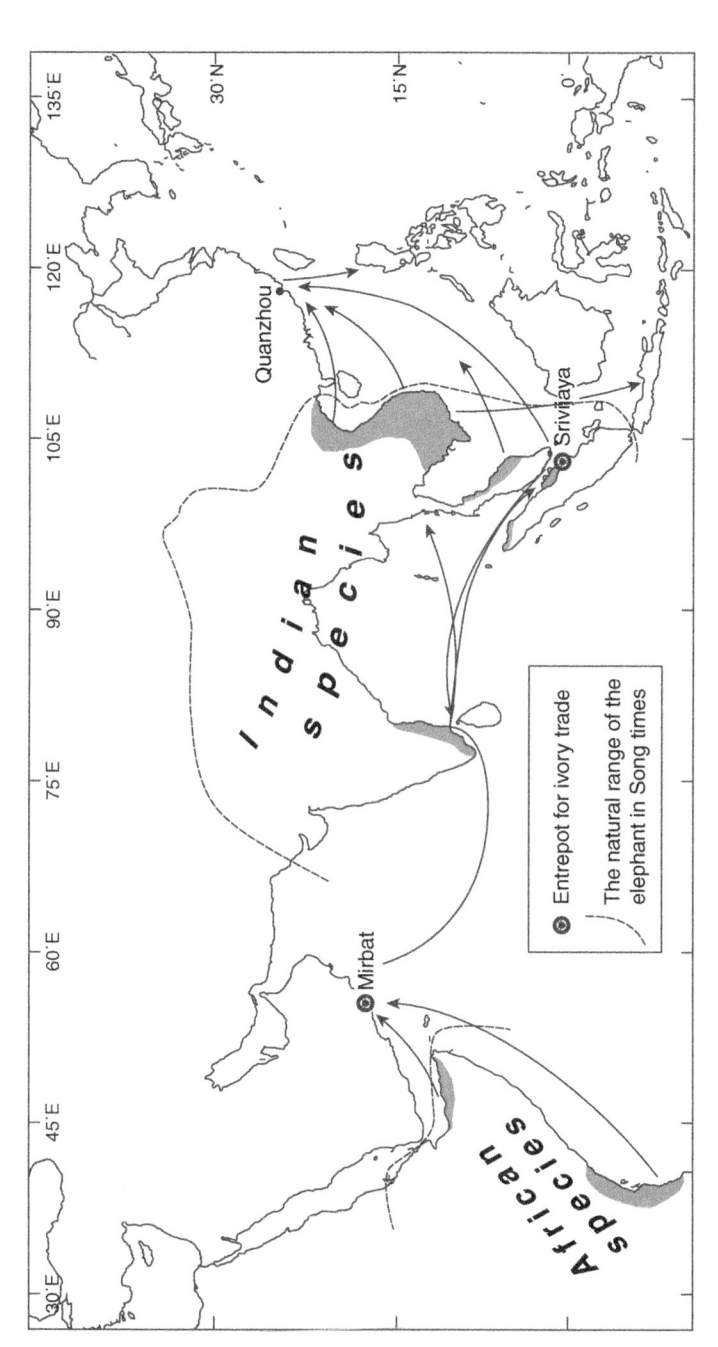

Figure 2.3 Sources of ivory and routes for the ivory trade during the Song Dynasty (adapted from Miksic, 1997: 28); map redrawn by David Quednau.

Figure 2.4 Elephant tusks from the *Java Sea Wreck* (photo © The Field Museum, Anthropology).

or sedative.[19] In contemporary Chinese medicine, ivory powder controversially continues to be sought out as a purgative and is believed to brighten the complexion.[20]

Ivory powder's therapeutic action on the liver would have been uniquely significant to the concerns of Song Dynasty women's medicine and to Song medical constructions of gender more broadly. In the *Synopsis of Formulas of the Golden Chamber*, Zhang prescribed long gu in his well-known formula *Gui Zhi Jia Long Gu Mu Li Tang* (Cinnamon Twig Decoction with Dragon Bone and Oyster Shell).[21] Ivory powder and other animal components of long gu worked in synergy with the botanical elements of the formula to correct sexual and energetic imbalance associated with nonreproductive desire, weakness, and dysregulation of sexual fluids.[22,23] Furth asserts that though human reproduction requires both sexes, the target of Song reproductive medicine was exclusively female.[24] The development of women's medicine in the forms of fuke and *chanke* (obstetrics) in Song Dynasty China had no correlating *nanke* (men's medicine) until modern times.[25] Furth writes that menstruation, as the "visible manifestation of the underlying functioning of the reproductive female body" became the object of intense Song medical scrutiny, discourse, and practice.[26] In the Yin-Yang opposition of *Xue* to *Qi* (Blood to Vital Energy), Furth states

Figure 2.5 A block of resin from the *Java Sea Wreck* (photo © The Field Museum, Anthropology, Catalog #351443).

that Blood in both its generalized, substantive yin aspect and as a "specifically female function" became a precious social "resource to be husbanded, nourished and used to contribute to human and cosmic creation."[27] Powdered ivory, in acting on the Liver as the storehouse of Blood, reified the sociopolitical and reproductive capacities of the Song elite.

Other medicinal objects from the *Java Sea Wreck* illuminate the multicultural context in which Chinese medicine developed in the Song and Yuan Dynasties, and the substantial influence of Persian and Arab medical knowledge on everyday practice. Eight blocks of resin (Figure 2.5) were recovered from the wreck, along with branches intended for use as aromatics or as dunnage.[28] Although the precise geographic sources and varieties of resin excavated from the *Java Sea Wreck* remain unidentified, multiple aromatic resins have been documented as trade goods along the Song and Yuan maritime trade routes. Further, many medicinal resins of Persian and Arab origin are listed in Zhao Rugua's thirteenth-century account of Chinese and Arab trade in the twelfth and thirteenth centuries.[29] Zhao's *Zhufan Zhi* (*Record of Foreign Peoples*) expounds:

> *Ju-hiang* [rǔxiāng, "milk incense," frankincense], or *hün-lu-hiang* (薰陸香) [Xūn lù xiāng], comes from the three Ta-shï [Arab/Persian] countries of Ma-lo-pa [southwestern Oman], Shï-ho [Yemen], and Nu-fa [southern Oman], from the depths of the remotest mountain valleys. The tree which yields this drug may, on the whole, be

compared to the *sung* (松 pine). Its trunk is notched with a hatchet, upon which the resin flows out, and when hardened, turns into incense, which is gathered and made into lumps. It is transported on elephants to the Ta-shï (on the coast); the Ta-shï load it upon their ships for barter against other goods in San-fo-ts'i [kingdom south of Quanzhou in the Southern Ocean, possibly Palembang on Sumatra]; and it is for this reason that the incense is commonly collected at San-fo-ts'i.[30]

The medicinal resins traded by Arabs and Persians, documented by Zhao Rugua in his capacity as inspector of foreign trade for Fujian, include camphor, frankincense, myrrh, dragon's blood, sweet benzoin, dammar, liquid storax, and benzoin.[31]

Ibn Sina, Chinese Herbs, and the *Hui Hui Yaofang*

In her work exploring the transfer of medicines from Iran and Arabia to China, Schottenhammer outlines medicinals of Persian and Arab origin prescribed in the thirteenth-century Chinese medical text, the *Hui Hui Yaofang* (*Islamic Prescriptions*). The *Hui Hui Yaofang* is a Yuan Dynasty compilation of Muslim medicine that closely resembles the Persian Ibn Sina's eleventh-century Arabic text, *al-Qānūn fī al-Ṭibb* (*The Canon of Medicine*).[32] In the century leading up to the publication of the *Hui Hui Yaofang*—the time frame during which the *Java Sea Wreck* vessel set sail—critical changes in the inventory of available drugs and to the clinical praxis of pharmacy took place in China.[33] According to Miyasita, this shift in Chinese materia medica was followed by substantial changes in Chinese diagnostics in the Yuan Dynasty.[34,35] Thought to be the official formulary of the Yuan administration, *Islamic Prescriptions* forever altered the catalog and practice of Chinese pharmaceuticals.[36] Within the mere three extant chapters of the formulary, 517 Islamic drugs, transliterated into Chinese from Persian and Arabic, are described.[37]

The rendering of the *Hui Hui Yaofang* from Ibn Sina's *Canon of Medicine* was undertaken by the director of the Yuan Medical and Pharmaceutical Bureau, a Nestorian Christian known as Isa Tarjaman.[38] Tarjaman established a capital hospital to formalize the practice of Islamic medicine in China, which he administered with his wife, Sara. Tarjaman also travelled from China to Maragha, the Persian capital of the Ilkhanate, center of scientific inquiry, and seat of the Church of the East.[39]

Ibn Sina's classic work was translated and diffused throughout the globe during the Middle Ages—in Europe under his Latinized name, "Avicenna"—influencing the practice of medicine in the East and the West. Translated into Latin in the twelfth century, Ibn Sina was established as a medical authority in Europe ranked with Hippocrates and Galen by the early fourteenth century.[40] His *Canon* synthesized works of Aristotle and Galen, according Galen central reverence as the root theorist of Islamic medicine. Beyond this, Ibn Sina summarized the two preceding centuries of Islamic medical knowledge and practice, drawing from the fruits of medical exchange across the vast borders of the Islamic world and including medical principles, anatomy, physiology, pathophysiology, disease prevention and treatment, materia medica, and exposition of the Galenic natural elements and bodily humors.[41]

Both the *Canon of Medicine* and the *Hui Hui Yaofang* extol the medicinal benefits of myrrh, which gained use in China as an analgesic aid to childbirth after its introduction during the Tang Dynasty.[42] The Ming *Bencao Gangmu* (*Compendium of Materia Medica*) indicates the use of myrrh for the treatment of wounds, miscarriage, and labor pains.[43] Imported and luxury aromatic resins seem to have gained a foothold in rituals surrounding childbirth in China's Middle Period (700–1300 CE), when the traditional management of labor was contested by emerging male chanke practitioners. Aromatic resins were burnt in ritual contexts by healers throughout Asia, but their use in officially regulated formulas dispensed by the new class of specialized physicians granted a new legitimacy and cachet to their usage.[44] According to Furth, Chen Ziming, the core Southern Song theorist of fanke and chanke, incorporated frankincense with other expensive aromatics alongside the powerful Chinese medicinal known as "winter solstice rabbit brain marrow" to hasten labor.[45] This juxtaposition allied luxurious imported medicinals, and the specialized knowledge of their use, with Chinese lunar fertility symbolism and Yin-Yang medical theory.[46]

Java Sea Wreck Ceramics

The majority of ceramics recovered from the *Java Sea Wreck* are utilitarian wares, such as green-glazed bowls likely from the Fujian kilns;[47] however, a significant number of pieces can be classified as high-quality ceramics that, based on stylistic and preliminary compositional analysis (the latter undertaken by one of the authors, Niziolek, at The Field Museum's Elemental Analysis Facility), were made at the famous kilns of Jingdezhen in Jiangxi province. These include finely made saucers, dishes, bowls, and ewers with a translucent blue *qingbai* (blue/green-white, pure) glaze. The ship was

Figure 2.6 *Kundika* from the *Java Sea Wreck* (photo © The Field Museum, Anthropology, Catalog #350819).

also carrying black- and brown-glazed tea bowls and Cizhou-type painted wares in the form of bowls, platters, ewers, bottles, and covered boxes. Although most of the trade ceramics were made in China for export, a few wares were identified as Southeast Asian.

Of the thousands of ceramic artifacts recovered from the *Java Sea Wreck* site, many have direct bearing on China's relationship to the Indian Ocean region's medical material culture during the Song and Yuan Dynasties. More than 350 Indian-style earthenware *kendis* and *kundikas* of unknown origin (possibly from southern Thailand) were excavated from the wreck, with *kundikas* (Figure 2.6) representing an older style of water vessel used in Hindu and Buddhist rituals.[48] *Kendis* (Figure 2.7) eventually became a popular item of trade for use in performing Islamic ablutions.[49] *Kendi* design evolved from the *kundika* ritual form in Southeast Asia, possibly in Indonesia, and was copied into Chinese design during the Tang Dynasty.[50] Wider-bodied kendis of the

Figure 2.7 *Kendi* from the *Java Sea Wreck* (photo © The Field Museum, Anthropology, Catalog #350715).

period retained religious use, but were eventually used in the preparation of brewed medicines.[51]

Also excavated with the wreck were five rare Indian-style ambrosia bottles, each measuring ten centimeters in height.[52] Identical in design to vessels containing the elixir of immortality held by deities in Indian religious artworks, the bottles have a distinctly phallic appearance and are suspected to have been for medicinal ritual use.[53] One of the bottles (Figure 2.8) is decorated with a largely effaced black underglaze cursive that resembles the Chinese character 子, *zi* (son, child, or seed). If the intended use of the ambrosia bottle was medical, this suggests a likely application of the object for fertility medicine. The juxtaposition of a Chinese inscription on an Indian-styled ritual medical object speaks to the presence

Figure 2.8 Ambrosia bottle from the *Java Sea Wreck* (photo © The Field Museum, Anthropology, Catalog #350273).

Exchanges and Transformations in Gendered Medicine 77

of syncretic or contesting cultural practices in the intended marketplaces to which the *Java Sea Wreck* vessel directed its course. Similar ambrosia bottles have been excavated from the *Breaker Shoal Shipwreck* in the Philippines.[54]

Ceramic Depictions of Botanics and Medicinals

Several of the kendi from the *Java Sea Wreck* collection are gourd-shaped (Figure 2.9). Though Indian-styled, they share with numerous Chinese ceramics excavated from the wreck an imitation of the natural forms of fruits and melons.[55] Given the use of these wares for the brewing, storage, and dispensing of medicines, their botanical-inspired designs and decorative motifs must be considered for their symbolic and ritual implications for Song Dynasty conceptions of health and fertility. Gourd-shaped, lobed ewers (Figure 2.10) were modeled in the Song period after metalware imported from Central Asia and the

Figure 2.9 Gourd-shaped *kendi* from the *Java Sea Wreck* (photo © The Field Museum, Anthropology, Catalog #350757).

Figure 2.10 Gourd-shaped qingbai ewer from the *Java Sea Wreck* (photo © The Field Museum, Anthropology, Catalog #350410).

Middle East.[56] Natural shapes, including lotus bowls, fruit-shaped jars and ewers, and melon-shaped boxes were popular trade items in the period and along with painted and etched floral and vegetal motifs often carried explicit reproductive and fertility imagery.[57]

The cross-pollination of ceramic forms, motifs, and technologies between China and the Near and Middle East during the Middle Ages has been well documented.[58] The multivocality of symbols deployed in motifs on medical ceramic wares assuredly created distinct but related meanings in the geographic and theoretical contexts on either side of the trade routes. Galenic, Islamic medicines and Song Dynasty fuke formulas would have been brewed and stored in nearly identical vessels and covered boxes across the region. In total, more than 250 sealed ceramic boxes (Figure 2.11), approximately 600 separate box lids, and 900

Figure 2.11 Qingbai-style covered boxes from the *Java Sea Wreck*. Designs starting from upper right and working clockwise: peony flower with wave border; possible strawberries; plain, peony flower; and stylized lotus spray (photo © The Field Museum, Anthropology).

additional box bases were excavated from the *Java Sea Wreck*.[59] Used for the storage of medicines, perfumes, mirrors, jewelry, and cosmetics,[60] many of the *Java Sea Wreck* boxes are inscribed with Chinese and Near and Middle Eastern text and decorative motifs. Small and miniature sizes of similar boxes have been documented for use as amulets.[61]

Box covers in the *Java Sea Wreck* collection span a wide variety of Chinese and Near and Middle Eastern decorative motifs. Chinese floral sprays (including peonies and lotus flowers) (Figure 2.11), leaping fish, boys-in-flowers, and inscribed Chinese text sit alongside Persian pomegranates (Figure 2.12), vegetal scrolls, boteh (Figure 2.13), Islamic basket-weave geometric motifs (Figure 2.14), and non-Chinese text. Many of the covered boxes are in the shape of lobed and ribbed melons (Figure 2.15), and one of the most intriguing boxes contains a small, sculpted, copulating couple (Figure 2.16).

As products of export and products consumed by the Song elite, the functionality of ceramics as decorative, domestic vehicles of emerging official Song reproductive medicines informs the domestic

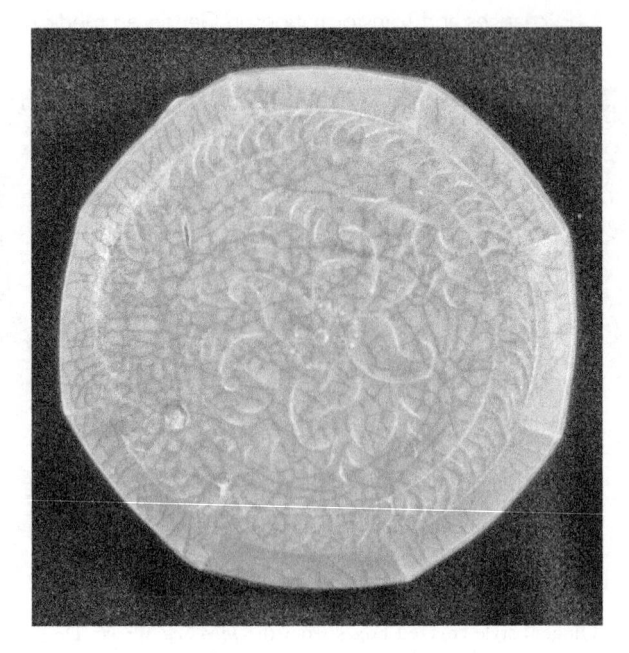

Figure 2.12 Covered box lid with pomegranate design from the *Java Sea Wreck* (photo © The Field Museum, Anthropology, Catalog #344638).

Figure 2.13 Covered box lid with boteh design from the *Java Sea Wreck* (photo © The Field Museum, Anthropology, Catalog #344300).

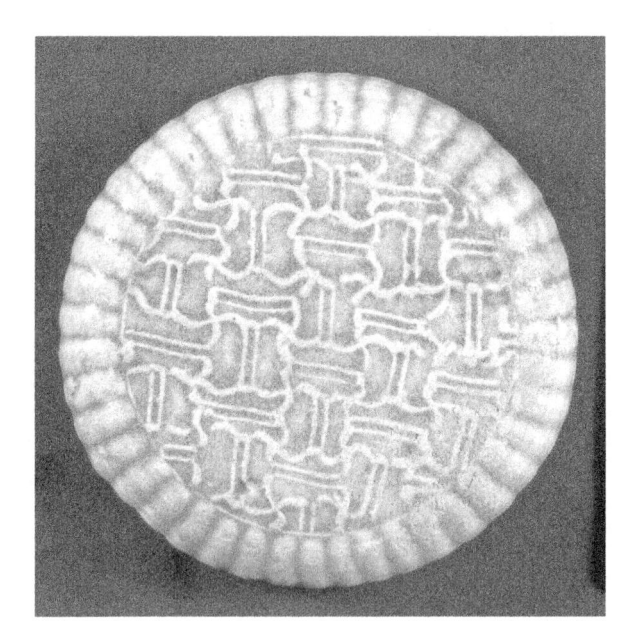

Figure 2.14 Qingbai-style covered box lid with basket weave design from the *Java Sea Wreck* (photo © The Field Museum, Anthropology, Catalog #344759).

Figure 2.15 Qingbai-style ribbed, melon-shaped box from the *Java Sea Wreck* (photo © The Field Museum, Anthropology, Catalog #344419).

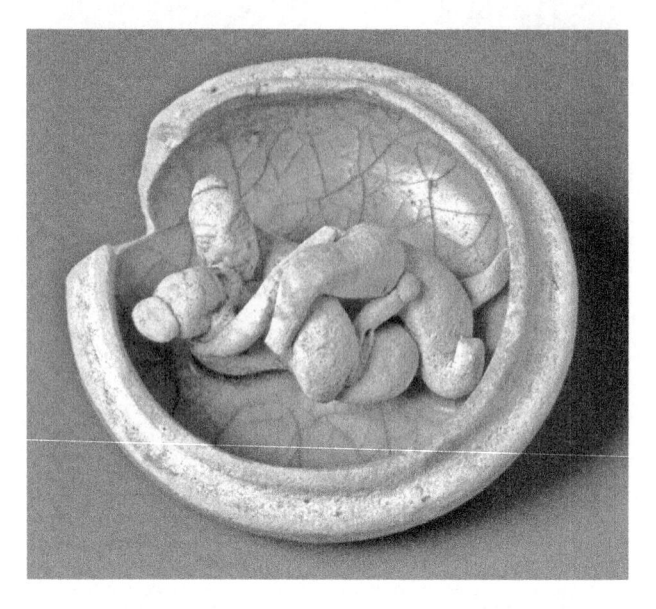

Figure 2.16 Qingbai-style covered box base with sculptural couple from the *Java Sea Wreck* (photo © The Field Museum, Anthropology, Catalog #344874).

habitus of household, gendered norms of sexuality and reproduction. As Furth notes, women's medicine was, after all, *wives' medicine*,[62] the fecundity of desire celebrated by these natural forms was medically normative only in the context of marital reproduction and domesticity. Song fuke texts are preoccupied with the application of emerging diagnostic theory and pharmaceutical praxis to the idealization of married women's menses.[63] Separate prescriptions for women emerged, linking the embodied consumption of medical formulas to the familial and reproductive cosmology of Yin-Yang theory.[64] Medical objects of household use were patterned after this cosmology in a manner that naturalized cosmological gender norms.[65]

Medical Symbolism and *Java Sea Wreck* Ceramics

Fruits

The majority of the Chinese and Islamic medical motifs found on ceramics from the *Java Sea Wreck* collection depict organic, botanical images of fruit and floral medicinals that loomed large in the symbolic-reproductive vocabularies of both cultural systems. The omnipresence of Song-style gourd- and melon-shaped ceramic wares (Figure 2.15) reflects a tendency in Song Dynasty art toward what Wirgin describes as "a realistic interpretation of nature, especially when a motif is recent and has not already obtained a traditionally established form."[66] According to Wirgin, melon motifs were not introduced to China until the tenth century and first appeared in Chinese art during the Song Dynasty.[67] Some varieties of melons and gourds were indigenous to China, but watermelon, known as *xi gua*, 西瓜 (western gourd), was an import from western Asia. According to Eberhard,[68] this transplant became an important symbol of female fertility. The *gua* character, 瓜 (melon/gourd), "can be divided down the middle" and with slight modifications each half can be read as the character *bā*, 八 (eight), an auspicious number because of its phonetic similarity to the word for wealth. The description of a newly matured girl as "twice times eight" or, referring to the red interior of a watermelon or womb, "as old as the divided gourd" indicated the onset of menses, sexual debut, and marriageability.[69] This explicit sexual symbolism is also found in the depiction of melons with a vine or stem: a pun phonetically and visually representing the wish for "ten thousand generations" of male descendants.[70] Eberhard also associates this fertility symbolism with the practice of young women picking melons, making offerings of gourds, and preparing melon cakes on special feast days and full moons so that they might be blessed by bearing sons.[71]

Melons and gourds, particularly their seeds, are thought to have pharmaceutical properties in Islamic humoral medicine. Galenic medicine views illness as a state of imbalance between four bodily elements, or humors—blood, phlegm, yellow bile, and black bile—with each humor being either hot or cold and dry or moist.[72] Ibn Sina asserted that the quality and balance of bodily humors, and therefore health, depended on the intake and digestion of appropriate foods, which were also divided into categories of hot or cold. Melons, being both cold and moist, helped achieve humoral balance and were capable of assuaging bodily heat, quenching thirst, cleansing the kidneys and the bladder, relieving burning sensations, and acting as a diuretic and pain reliever.[73] Chinese medicine put transplanted melons to similar use. Melon stalk powder, *Gua Di San*

(瓜蒂散), became a powerful diffusing and ejecting formula used to clear phlegm and eject cold evil from the body.[74]

Another fruit not native to China that appears on *Java Sea Wreck* ceramic motifs is the stylized pomegranate (Figure 2.12). Transplanted to China from the Near East during the Han Dynasty, the pomegranate carries sexual symbolism similar to the watermelon in both locations.[75] Because their interiors are both red and full of seeds, both fruits symbolized prolific female fertility in their new Chinese semiotic soil.[76] Objects depicting pomegranates became a popular choice for a wedding gift, often inscribed with the blessing "hundred seeds, hundred sons."[77,78]

Pomegranate fruits had long been a potent fertility symbol in pre-Islamic Iran, and had been brought under the banner of Islamic medicine in the works of the Shiite Imam Ja'far al-Sadiq.[79] Describing the fecundity, sexuality, and desire inherent in Persian pomegranate symbolism, Mottahedeh writes:

> The pleasures embodied in this fruit are rooted firmly in the affects and in the delights of comely maidens in paradise, yet they also blossom in the exuberance of the much feared and desiring bodies of earthly women. Capable of "bleeding" the pomegranate embodies, in ancient mythologies, the force of life and death itself, and therefore appears often as a symbol of the mother's menstruating and fertile body.[80]

This imagery of blood as the source of female fertility carried profound resonance with Song Dynasty fuke. An astringent, digestive aid, and antiparasitic,[81] pomegranate was adopted into Chinese materia medica and ceramic design. Pomegranate motifs retained their Islamic functions as reminders of God's earthly mercies and seductive indicators of a Heavenly Paradise to come[82] while doing double-duty as mirrors of elite Song reproductive womanhood. In China, these designs frequently featured exaggerated sepals and calyxes with composite lotus and peony flowers integrated into a new and hybrid plant.[83]

Flowers

Peony flowers appear on numerous ceramic wares and covered boxes from the *Java Sea Wreck* cargo (Figure 2.11). Symbolizing wealth and distinction as "the queen of flowers," according to Eberhard, the peony also symbolized a beautiful woman whose perfume attracts men, or more explicitly, indicated receptive female genitalia.[84] Eberhard writes, "When the dew (= semen) drops, the peony opens."[85] This imagery on covered boxes used to store medicines and cosmetics would have unmistakable relevance for elite Song wives. In the peony's explicit capacity as a female sexual symbol, it was used medicinally to address Yin-Yang imbalance by constraining yin when it was out of bounds.[86] Peony was also used for depressed Liver Qi and construction-Blood disharmony, breast and menstrual pain and irregularities, emotional imbalance, digestive ailments, umbilical pain, to clear organ heat and to nourish the all-important blood.[87] Peony was used in Islamic formularies as an anti-inflammatory, antispasmodic, heart tonic, and antihysteria drug.[88]

The lotus plant is another floral medicinal featured prominently on the *Java Sea Wreck* ceramic wares (Figure 2.11). Lotus embryo was used medicinally to clear heat, to treat "upflaming heart fire," and to treat heart and kidney yin vacuity, common causes of infertility.[89] Lotus fruits were also used as astringents, to nourish the heart, and to quiet the spirit, while lotus leaves were used to staunch excessive bleeding.[90] Symbolizing purity and rebirth in Buddhism, the lotus carried additional symbolism in China based on the phonetics of its name. Called both *hé* (荷) or *lián* (莲), hé indicated concord or unison while lián means to bind or connect, as in marriage, to be uninterrupted, or modesty.[91] Depending on the symbolic context and audience, a lotus flower could symbolize uninterrupted unity and spiritual rebirth and/or a modest wife. Similar to other organic motifs discussed here, the lotus blossom symbolized the female yin aspect of creation, and the stem the male yang aspect.[92]

A unique hybrid-floral symbol is etched on the lid of one of the *Java Sea Wreck's* boxes (Figure 2.13). A floral spray combining lotus at the base, a stylized, exaggerated lotus-like pomegranate calyx with four paisley sepals sits at the center, and a distinctive Persian boteh flower curls above. Boteh motifs are a hallmark of Persian art, appearing in pre-Islamic Iran as the wings of Homa and Senmurv and transforming, according to Cyrus Parham, in the early centuries of Islam into a stylized cypress symbolizing "immortality and timelessness."[93] *Boteh* is the Persian word for a shrub, thicket, herb, and sometimes leaf or flower.[94] Known as *paisley* in the West, the boteh is distinctively shaped like a teardrop with a curved, pointed tip and elaborate border surrounding its edge, often arranged into stylized depictions of oversized botanicals.[95,96] Robinson describes the essential symbolism of boteh as a "seed pod unit" representing regeneration, life force, and fertility.[97]

Boteh motifs can be simple or ornate, and came to dominate Persian design in the tenth century. At times seeming to represent an almond, a pear or cut fig, a Zoroastrian flame, or even a Bodhi leaf, the motif eventually spread across western Asia.[98,99,100] A twin boteh, with two units sitting side by side, appears like a stylized blossom in Iran and Egypt at this time, with Iranian architectural panels excavated from the Sabz Pushan site at Nishapur (an important junction on the Silk Road) identical to the Yin-Yang configuration of the Chinese Taiji symbol.[101,102] Many bowls in the *Java Sea Wreck* collection feature etched flowers on the circular base. A small sample of these stylized blossoms strongly resemble the Taiji symbol, with floral lines standing in for the boundary between yin and yang (Figure 2.17).

Figure 2.17 Bowl with design similar to the Taiji symbol in the center from the *Java Sea Wreck* (photo © The Field Museum, Anthropology, Catalog #345954).

Phallic Designs, Vegetal Scrolls

Multiple covered boxes from the shipwreck are inscribed on the underside of their bases, hidden from view. Many of these hidden inscriptions refer to family names, production sites, or are simply numeric. Similar box bases with hidden inscriptions have been noted in China, often referring to the contents of a particular box, either a mirror or the name of a particular medicine.[103] Several of the hidden inscriptions stand out in the *Java Sea Wreck* collection because of their possible relation to medical practices, specifically those related to fertility. One qingbai box base (Figure 2.18) has been found with a design resembling an upright phallus from which an exuberant Near Eastern-style vegetal scroll emerges.[104] Given the context of similar covered boxes' use in the storage of medicines, it is highly likely this box was intended for a fertility medicine or marital gift.

Figure 2.18 Qingbai-style covered box base with molded phallic design (highlighted in black) from the *Java Sea Wreck* (photo © The Field Museum, Anthropology, Catalog #345128).

The symbolism in this piece can be read multivocally, first, for the meanings associated with this type of vegetal scrollwork in Islamic art. Alternately, it can be read for the similarity in the depiction of an exposed phallus among vegetation in similar Chinese motifs. It is important to note that the terms Islamic art and Islamic medicine, both expressions of Islamic civilization, do not always indicate an Arab or even a religious origin. Though many innovations in both fields are explicitly Islamic, many more are pre-Islamic phenomena created by the diverse cultures making up the so-called Islamic world.[105] Oweis defines the scope of Islamic art by identifying its shared visual elements across cultures: calligraphy, geometric patterns, and Arabesque.[106] The vegetal scrolls seen on the *Java Sea Wreck* object in Figure 2.18 resemble the Arabesque motif of nonrepresentational, interlacing botanical scrolls.[107] Arabesque vegetal scrolls originate in pre-Islamic depictions of the stems of lush, fertile plant life prolifically splitting off into interlacing lines, multiplying, and turning back to their original source.[108,109] The emergence of an Arabesque vegetal scroll from an erect phallus recalls Chinese images of Indian-styled figures with lotus flowers blossoming from the navel.[110] Common in the Song Dynasty and thereafter, images of Indian-adorned (or semi-adorned, most figures lacked clothing below the waist) male children among floral scrolls symbolized prolific reproduction and rebirth.[111] Both motifs feature exposed penises among vegetation,[112] though the disembodied and concealed phallus on the *Java Sea Wreck* artifact is unique.

Geometric Designs

A single covered box lid from the *Java Sea Wreck* collection is decorated with a distinctive geometric, interlaced basket weave motif (Figure 2.14). Geometric basket weave motifs are a variation on interlace and lattice Islamic design vocabularies often seen in the art and architecture of the Muslim world.[113] In describing the principles of Islamic design, Oweis identifies the repetition of a single unit or cell and symmetrical, balanced proportions as indicative of an underlying cultural view of geometry—what Abas and Salman call "the unifying intermediary between the material and the spiritual world."[114] The symbolism of this geometry allows a visual discourse on living beings disallowed by taboos on directly representational art. Critchlow expands on the geometric cosmology of

Islam by delineating the three "most fundamental figures of Islamic art," the triangle, the square, and the hexagon.[115] Written into these forms are the tenets of Galenic medicine: from the circle comes the triangle, one point expanded to three, symbolizing human consciousness made material by the three basic biological functions: "ingestion, digestion, and excretion."[116] This depiction of "life without figural form" can be broadened by expanding the circle once again: the four-pointed square repeated in the basket weave motif symbolizes the materiality of the earth as a whole, with each point symbolizing the four Galenic elements of the universe: Water, Earth, Fire, and Air.[117,118] El-Said and Parman write that this geometry is the "unifying concept of composition" in Islamic design, "despite the diversity of materials, forms, or styles used."[119]

Wu Xing (Five Phases)

Finally, a handful of *Java Sea Wreck* inscriptions etched on the underside of qingbai covered box bases appear to contain a Chinese radical essential to Song Dynasty women's medicine: 木 *Mù* (wood/tree) (Figure 2.19). In the Chinese *Wu Xing* (Five Phases) system of correspondences, Wood connotes the organic qualities of growth and life that are reborn in the spring after a long winter and that underlay the progression of conception, birth, and new life. Porkert argues that Chinese medical thinking, and its Five Phases (often misleadingly translated as "Five Elements") is as "heavily weighted toward function as that of Galen toward structure."[120] Five Phase functional philosophy was first documented in China in the fourth century BCE, and was systematized by Zou Yen and his followers sometime between 350 and 270 BCE.[121] It describes the five quintessential processes or phases that animate all living entities, from a solitary body to a larger ecosystem. Wu Xing literally translates to five walks, circulations, or movements, and indicates process and relationship instead of static material objectification. In other words, the "elements" are not basic units of matter as in Galen; they are basic qualities of action.[122]

The Five Phases are Wood, symbolizing spring and new growth; Fire, symbolizing summer and flowering; Earth, symbolizing late summer and fruition; Metal, symbolizing autumn and harvest; and Water, symbolizing winter, retreat, and stillness. The Five Phases also define a system of classification and correspondences for all phenomena, with each phase denoting a particular color, time of day, shape, feature of the landscape, internal organ, orifice, direction, climate, sound, and taste.[123] Five Phase philosophy was a critical part of the timing and performance of ritual, the practice of medical diagnostics and treatment, and the creation of works of art.[124] The Five Phases cyclically generate and constrain one another, and harmonious balance between them results in bodily health.

Writing about the relationship between the Wood phase and reproduction, Furth describes the medical teaching of Yuan Dynasty physician Zhu Zhenheng: "Ministerial Fire lodged in the Liver system is the source of sexual desires in both males and females...This is further demonstrated by the cosmological affinities between Liver, the phase of Wood and the mating season of spring."[125] Song and Yuan period fuke medicines prioritized the Wood phase and its corresponding Liver system as the storehouse of Blood. Yang Shiying, a Southern Song medical authority from Fujian, the province of the probable port of departure of the *Java Sea Wreck* vessel, wrote: "In women blood is fundamental. Why? Because their Blood is in ascendancy over qi; it is stored in the Liver system, flows through the womb and is ruled by the Heart system; it ascends to become breast milk, descends to become menses, unites with semen to make the embryo."[126] Medicinals with action on the body's Wood phase, Blood, and Liver therefore proliferated in the Song and Yuan periods.

The meaning and origin of the bottom half of the pictured inscriptions, an oval shape emerging from a horizontal line, is unknown. In one example, the symbol emerges like a cross from a flower (Figure 2.19). This symbolism of a cross rooted in a flower has profound significance for the Fujian port community of Quanzhou, from where the *Java Sea Wreck* vessel is believed to have set sail. Multiple cross-on-flower gravestone inscriptions from the period mark Nestorian Christian burials in Quanzhou.[127] The Nestorian community in Quanzhou made up a part of the multicultural city of merchants and artisans known by the Arabic term *Zaytun*.[128] Nestorian speakers of Syriac, Persian, and Arabic—like Isa Tarjaman and his wife—played an important role in the dissemination and development of medical knowledge throughout the Near and Middle East and China, including the translation and compilation of the *HuiHui Yaofang*.[129] The resemblance borne by the hybrid Five Phase Wood/cross-on-flower symbolism on covered

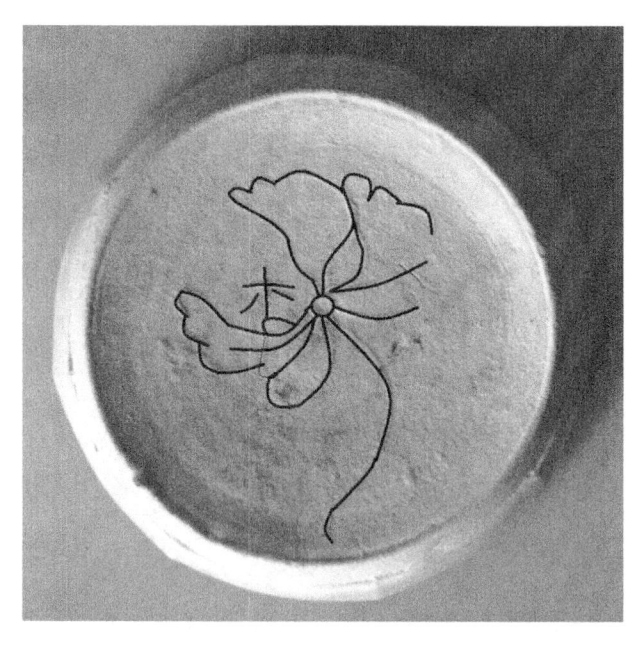

Figure 2.19 Qingbai-style covered box base with molded flower and inscription (highlighted in black) from the *Java Sea Wreck* (photo © The Field Museum, Anthropology, Catalog #345136).

storage boxes from the *Java Sea Wreck* to inscriptions on the tombstones of Zaytun merits further investigation.

Numeric

Finally, numeric symbols found on covered box bases from the *Java Sea Wreck* cargo bear testimony to the emerging culture of science and medicine in the thirteenth-century IOW. Five numeric systems are represented in inscriptions from the shipwreck, including what appear to be examples of the early spread of Eastern Arabic-Indic numerals to East and Southeast Asia. This number system emerged from the unique intercultural exchanges between South Asia and the Muslim World across regional trade routes, spread to the far corners of the globe, and has since transformed and come to dominate global math, science, medicine, and technology as their symbolic lingua franca.[130] An inscription highlighted in black from the *Java Sea Wreck* resembles what appears to be the Eastern Arabic (Arabic-Indic) number ٦ *Shesh* (6) (Figure 2.20).

Figure 2.20 Qingbai-style covered box base with inscription (highlighted in black) from the *Java Sea Wreck* (photo © The Field Museum, Anthropology, Catalog #345116).

Conclusion

A selection of artifacts from The Field Museum's *Java Sea Wreck* collection represent a cross-section of medical goods exchanged along the maritime trade routes linking East Africa, the Middle East, and South, East, and Southeast Asia in the thirteenth century. An analysis of these objects reveals a dialogue between Song and Yuan period Chinese medicine with medical knowledge and medicinals originating from across the IOW. At the dawn of the specialization of women's medicine in China, this network of foreign trade informed the practice of medicine and the construction of gender among the elite. Though styled discursively as homogeneous and traditional, the classical Chinese medicine that emerged in this period incorporated significant medical knowledge and material culture derived from Near Eastern and Islamic medicine.

This chapter scratches the surface of the medical implications of the artifacts housed in the *Java Sea Wreck* collection. Further exploration of this facet of the collection can shed additional light on the impact of the cultural exchanges along the medieval Maritime Silk Road on the global systems of medicine along its borders.

Notes

1. Macao Museum (2012), *Maritime Porcelain Road—Relics from Guangdong, Hong Kong and Macao Museums* (Macao: Macao S. A. R. Cultural Affairs Bureau and the Macao Museum Binding).
2. M. Flecker and W. M. Mathers, eds. (1997), *Java Sea Wreck Archeological Report* (Annapolis: Pacific Sea Resources), pp. 1, 20, 77.
3. C. Furth (1999), *A Flourishing Yin: Gender in China's Medical History* (Berkeley: University of California Press), p. 66.
4. Ibid., p. 60.
5. Z. Wang, P. Chen, and X. Xing (1999), *History and Development of Traditional Chinese Medicine* (Beijing: Science Press), p. 72.
6. J. Needham (1956), *Science and Civilization in China,* vol. 2, History of Scientific Thought (Cambridge: Cambridge University Press), p. 56.
7. N. Sivin (1988), "Science and Medicine in Imperial China—The State of the Field," *The Journal of Asian Studies,* 47 (1): 47.
8. I. Robinet, P. Wissing, trans. (1990), "The Place and Meaning of the Notion of Taiji in Taoist Sources Prior to the Ming Dynasty," *History of Religions,* 29 (4): 380.
9. N. Wiseman and Y. Feng (1998), *A Practical Dictionary of Chinese Medicine* (Brookline, MA: Paradigm), p. 705.
10. N. Wiseman and A. Ellis (1996), *Fundamentals of Chinese Medicine* (Brookline, MA: Paradigm), pp. 2–5.
11. Flecker and Mathers, eds., *Archeological Report.*
12. M. Flecker (2011), "Wrecked Twice: Shipwrecks as a Cultural Resource in Southeast Asia," in J. N. Miksic, G. Y. Goh, and S. O'Connor, eds., *Rethinking Cultural Resource Management in Southeast Asia: Preservation, Development, and Neglect (*London: Anthem Press).
13. Flecker and Mathers, eds., *Archeological Report.*
14. Ibid.
15. Ibid.
16. Ibid., pp. 28–29, 81; R. Zhao Friedrich Hirth and William Woodville Rockhill, trans. (2012 [1911] [circa 1200]), *Zhufan Zhi. Chau Ju-Kua: His Work on the Chinese and Arab Trade in the Twelfth and Thirteenth Centuries, Entitled Chu-fan-chï* (Hong Kong: Forgotten Books), p. 232.
17. B. Laufer (1925), *Ivory in China, Anthropology Leaflet 21* (Chicago, IL: Field Museum of Natural History), pp. 16, 25–26.

18. J. N. Miksic (1997), "Historical Background," in Flecker and Mathers, *Archeological Report*, p. 14.
19. B. Laufer, *Ivory in China*, p. 26.
20. D. Levin (March 1, 2013), "An Illicit Trail of African Ivory to China," New York Times, http://www.nytimes.com/2013/03/02/world/asia/an-illicit-trail-of-african-ivory-to-china.html?pagewanted=all&_r=0, accessed March 2, 2013.
21. C. Liu and P. Chen (1999), *Well-Known Formulas and Modified Applications* (Beijing: Science Press), p. 33.
22. Furth, *Flourishing Yin*, pp. 15, 89.
23. Wang, Chen, and Xing, *History and Development*, pp. 33–34.
24. Furth, *Flourishing Yin*, p. 74.
25. Ibid., p. 92.
26. Ibid., p. 66.
27. Ibid., p. 93.
28. Flecker and Mathers, eds., *Archeological Report*, p. 81.
29. Zhao, *Zhufan Zhi*, pp. 193–202.
30. Ibid., p. 195.
31. Ibid., pp. 43, 193–202.
32. A. Schottenhammer (2010), "Transfer of *Xiangyao* 香藥 from Iran and Arabia to China—A Reinvestigation of Entries in the *Youyang zazu* 酉阳杂俎 (863)," in *Aspects of the Maritime Silk Road: From the Persian Gulf to the East China Sea*, ed. Ralph Kauz (Weisbaden: Harrassowitz Verlag), p. 128.
33. S. Miyasita (1976), "A Historical Study of Chinese Drugs for the Treatment of Jaundice," *American Journal of Chinese Medicine*, 4 (3): 239.
34. Ibid.
35. Sivin, "Science and Medicine," 47 (1): 69.
36. Y. C. Kong and D. S. Chen (1996), "Elucidation of Islamic Drugs in Hui Hui Yaofang: A Linguistic and Pharmaceutical Approach," *Journal of Ethnopharmacology*, 54 (2–3): 85.
37. Ibid.
38. H. Selin, ed. (1997), *Encyclopedia of the History of Science, Technology, and Medicine in Non-Western Cultures* (Dordrecht: Kluwer), p. 454.
39. Ibid.
40. U. Weisser (1987), "Avicenna xiii: The Influence of Avicenna on Medical Studies in the West," *Encyclopaedia Iranica*, 3 (1):107.
41. B. Musallam (1987), "Avicenna x. Medicine and Biology," *Encyclopaedia Iranica*, 3 (1): 94–96.
42. Li Shizen (1990 [1578]), *Bencao gangmu*, 2nd ed. (Beijing: Renmin weisheng chubanshe), p. 30.
43. Ibid.
44. Flecker and Mathers (eds.), *Archeological Report*, p. 29.
45. Furth, *flourishing yin*, p. 177.
46. Ibid.
47. Brown, "Ceramics inventory."

48. Ibid., p. 172.
49. Macao Museum, *Maritime Porcelain Road*.
50. Brown, "Ceramics inventory,: p. 172.
51. Ibid.
52. Ibid., p. 130.
53. Ibid.
54. M. Dupoizat (1995), "The Ceramic Cargo of a Song Dynasty Junk Found in the Philippines and Its Significance in the China-South East Asia Trade," in *South East Asia and China: Art, Interaction and Commerce* ed. R. Scott (London: Percival David Foundation of Chinese Art), pp. 205–224.
55. R. Brown, "Ceramics Inventory," p. 174.
56. D. P. Leidy, W. F. A. Siu, and James C. Y. Watt (1997), "Chinese Decorative Arts," *The Metropolitan Museum of Art Summer Bulletin*, 55 (1): 5.
57. L. Rotondo-McCord (2001), *Heaven and Earth Seen Within: Song Ceramics from the Robert Barron Collection* (Jackson: University Press of Mississippi), p. 17.
58. P. Morgan (1991), "New Thoughts on Old Hormuz: Chinese Ceramics in the Hormuz Region in the Thirteenth and Fourteenth Centuries," *Iran*, 29: 67.
59. Brown, "Ceramics inventory," p. 133.
60. Ibid., p. 131.
61. Ibid., p. 108.
62. Furth, *Flourishing Yin*.
63. Ibid., p. 76.
64. Ibid., pp. 24, 85.
65. Ibid., p. 25
66. J. Wirgin (1970), *Sung Ceramic Designs* (Stockholm: Elanders Boktryckeri Aktiebolag), p. 179.
67. Ibid.
68. W. Eberhard (1986), *A Dictionary of Chinese Symbols: Hidden Symbols in Chinese Life and Thought* (London: Routledge & Kegan Paul).
69. Ibid., pp. 132–133.
70. S. Cammann (1962), *Substance and Symbol in Chinese Toggles: Chinese Belt Toggles from the C.F. Beiber Collection* (Philadelphia: University of Pennsylvania Press), p. 115.
71. Eberhard, *Dictionary of Chinese Symbols*, pp. 132.
72. A. A. Afkhami (2004), "Humoralism (or Galenism)," *Encyclopaedia Iranica*, 12 (6): 566–570, p. 566.
73. H. A'lam (1993), "Cucumber," *Encyclopaedia Iranica*, 6 (5): 450–452.
74. Wiseman and Feng, *A Practical Dictionary*, pp. 128, 170, 254, 743.
75. Wirgin, *Sung Ceramic D*, p. 177.
76. Eberhard, *Dictionary of Chinese Symbols*, p. 240.
77. Ibid., pp. 240–241.
78. Wirgin, *Sung Ceramic Designs*, p. 177.
79. A. K. Moussavi (2008), "Ja'far al-Sadeq v. and herbal medicine," *Encyclopaedia Iranica*, 14 (4): 264.

80. N. Mottahedeh (2008), *Representing the Unpresentable: Historical Images of National Reform from the Qajars to the Islamic Republic of Iran* (Syracuse: Syracuse University Press), pp. 45–46.
81. N. Wiseman and Y. Feng (1998), *A Practical Dictionary of Chinese Medicine* (Brookline, MA: Paradigm), pp. 183, 298, 318, 518, 445.
82. F. S. Oweis (2002), "Islamic Art as an Educational Tool about the Teaching of Islam," *Art Education*, 55 (2): 23.
83. Wirgin, *Sung Ceramic Designs*, pp. 177–178.
84. W. Eberhard, *Dictionary of Chinese Symbols*, pp. 231–232.
85. Ibid., pp. 255.
86. Wiseman and Feng, *A Practical Dictionary*, p. 255.
87. Ibid., pp. 110, 426, 493, 731.
88. F. Ahmad, N. Tabassum, and S. Rasool (2012), "Medicinal Uses and Phytoconstituents of *Paeonia officinalis*," *International Research Journal of Pharmacy*, 3 (4): 85–86.
89. Wiseman and Feng, *A Practical Dictionary*, p. 263.
90. Ibid., pp. 490, 518, 569.
91. Eberhard, *Dictionary of Chinese Symbols*, pp. 168–170.
92. Ibid., p. 170.
93. C. Parham (1999), "From Cyprus Tree to Botteh (Paisley)," *Nashr-e Danesh*, 16 (4): 1378, trans. Shahin Shahin, http://www.rugidea.com/from_cypress_tree_to_botteh.html, accessed August 20, 2013.
94. K. E. Eduljee (2007), "Boteh," *Zoroastrian Heritage*. http://www.heritage-institute.com/zoroastrianism/trade/paisley.htm, accessed February 3, 2013.
95. Ibid.
96. M. Maskiell (2002), "Consuming Kashmir: Shawls and Empires, 1500-2000," *Journal of World History*, 13 (1): 29.
97. S. Robinson (1969), *A History of Printed Textiles* (Cambridge: MIT Press), p. 115.
98. Parham, "Cyprus Tree to Botteh."
99. Eduljee, "Boteh."
100. P. Cummings (2008), "Paisley: A Brief History," *Quilter's Muse Publications*, http://www.quiltersmuse.com/Paisley-A-Brief-History.html, accessed August 20, 2013.
101. Eduljee, "Boteh."
102. M. Sardar (2001), "The Metropolitan Museum's Excavations at Nishapur," *Heilbrunn Timeline of Art History* (New York: The Metropolitan Museum of Art, 2000–2001), http://www.metmuseum.org/toah/hd/nish/hd_nish.htm, accessed February 3, 2013.
103. Brown, "Ceramics Inventory," p. 137.
104. Maura Condon, Electronic Communication, August 16, 2012.
105. O. Graber (1987), *The Formation of Islamic Art* (New Haven, CT: Yale University Press), p. 2.
106. Oweis, "Islamic Art," 19.
107. Ibid.
108. Ibid., 23.

109. D. Jones (1978), "Surface, Pattern, and Light," in *Architecture of the Islamic World: Its History and Social Meaning*, ed. G Michell (London: Thames and Hudson), p. 171.
110. Wirgin, *Sung Ceramic Designs*, p. 180.
111. Ibid., pp. 179–180.
112. Wirgin, *Sung ceramic designs*, p. 183.
113. L. Prasertwaitaya (2007), *Untangling the Arabesque: Islamic Design Elements in the Monroe Park Campus*, www.library.vcu.edu/exhibits/documents/UntanglingtheArabesque2.pdf, accessed February 3, 2013, p. 88.
114. S. J. Abas and A. S. Salman (1995), *Symmetries of Islamic Geometrical Patterns* (Singapore: World Scientific), p. 9.
115. K. Critchlow (1976), *Islamic Patterns: An Analytical and Cosmological Approach* (Rochester: Inner Traditions), p. 16.
116. Ibid., p. 16.
117. Jones, "Surface, Pattern, and Light," p. 23.
118. S. H. Nasr (1987), *Islamic Art and Spirituality* (Albany: SUNY Press_.
119. I. El-Said and A. Parman (1976), *Geometric Concepts in Islamic Art* (Palo Alto, CA: Dale Seymour), p. xi.
120. M. Porkert (1974), *The Theoretical Foundations of Chinese Medicine: Systems of Correspondence* (Cambridge: MIT Press), pp. 107–108.
121. J. Needham (1969), *The Grand Titration: Science and Society in East and West* (Cambridge: Cambridge University Press), pp. 232, 242.
122. Needham, *Science and Civilization*, vol. 2, pp. 243–268.
123. M. Maruyama (1952), *Studies of the Classics of Acupuncture Medicine* (Osaka: Sogen), pp. 15–25.
124. Needham, *Science and Civilization*, vol. 2, p. 231.
125. Furth, *Flourishing Yin*, p. 171.
126. Ibid., p. 73.
127. S. N. C. Lieu (2009), "Manichaean and Nestorian Remains in Zayton (Quanzhou)," in *New Light on Manichaeism: Papers from the Sixth International Congress*, ed. J D Duhn (Leiden: Brill), pp. 194–195; p. 194.
128. Ibid., pp. 194–195.
129. Selin, *Encyclopedia of the History of Science*.
130. E. C. Bayley (1883), "On the Genealogy of Modern Numerals: Part II, Simplification of the Ancient Indian Numeration," *The Journal of the Royal Asiatic Society of Great Britain and Ireland, New Series* 15 (2):1–72.

3

Saints, Serpents, and Terrifying Goddesses: Fertility Culture on the Malabar Coast (c. 1500–1800)

P. K. Yasser Arafath

This chapter examines the unique and interlocking dynamics of fertility rituals and practices of different cultural traditions in the Malabar region, on the western coastal area of Kerala, between the sixteenth and the nineteenth centuries. Healing traditions in this region reflected the plurality and interdependence of different cultural frameworks that contributed to the range of medical knowledge in this period. Concepts of illness, including those relating to fertility issues, were constructed through the religious-cultural consciousness of the people. This chapter tries to explore some areas of what may be termed fertility psyche, beliefs and practices, in relation to the binding factors of religion, social functioning, caste institutions, and various concepts such as pollution. Although humor-based healing traditions such as Ayurveda coexisted with magico-spiritual practices, the belief in the supernatural was very predominant in the region. These beliefs created a large complex of rituals and festivals that also created a network of fertility geographies and cults across the Malabar region. These constituted a parallel world of alternative and non-textual fertility healing practices.

These centuries witnessed the gradual crystallization of fertility norms and cults across religious communities, and people attributed issues such as female fertility and male impotence to a multitude of gods, mother goddesses, holy men, and certain animals. Thus, the sacred places these forces were meant to reside in, such as temples, churches, mosque, *kavus* (groves considered sacred by Hindus), and *dargahs*, created a constellation of fertility complexes across the Malabar region. Each sociocultural group developed their own understandings of fertility practices, and matters such as conception, pregnancy, and childbirth were located within a larger constellation of beliefs, knowledge traditions, experiences, facilities, and resources. Malabar, being a cultural melting pot during the medieval and early modern period in the Indian Ocean world (IOW), became the site for the evolution of an interesting fertility psyche. This was centered, in particular, on concepts such as hygiene and pollution that affected the body—both individual and social. Fertility beliefs in all religious communities ensured multiple boundaries for desire, sensuality, sexuality, and bodily engagements. The fertility networks that evolved through these boundaries subsequently shaped multiple modes of ordering—of movements, caste relations, and religious practices.

By studying the fertility cultures of the Malabar region, a complex social, religious, and textual history can be identified from the sixteenth to the nineteenth centuries. This chapter draws on a wide range of language sources: Malayalam, Arabic, and Arabic-Malayalam, as well as a large number of translated travelogues and a part of the colonial archive. I believe that few of the Arabic and Arabic-Malayalam texts utilized in this chapter, such as the *Fath-hul-Muin, Nafesathmala, Muhiyuddheenmala,* and the *Fee Shifau-n-Nasi*, have been used in the existing writings on the history of Malabar. Considering these texts will enable a preliminary examination of the ways in which Malabar was influenced by the transnational network of Muslim scholars across what may be termed as the shafiite knowledge grid in the IOW over a period of three hundred years.

Fertility and "Terrifying" Goddesses

Among the mighty fertility agents, enormously powerful and sexually aggressive virgin mother goddesses, known collectively as "Bhagavatis," created a major fertility psyche in the region. These goddesses, at places such as the Kodungallor and Chottanikkara temples, trace their origin to the early medieval period. They were worshipped by pregnant women in order to ward off, among other things, the evil influences of smallpox (*Masuri*), chicken pox (*Ponganpani*), and bodily possession. In the sixteenth century, people converged at Chottanikkara for the removal of *unmadam* and other serious psychological problems.[1] Female believers also invoked the mother goddesses to enhance their fertility.[2] Cultural and liturgical rituals associated with these popular goddesses, who were known for their divine virility and nearness, created a psychological togetherness in believers. The excessive social stigma attached to barrenness and female infertility was one of the main reasons why women resorted to the divine intervention of goddesses such as the Kodungallur Bhagavati. *Komarams* (oracles or the dancing priests) at these temples acted as the divine interlocutors between the fertility deities and female worshipers. Known for their agency in performing miracles and their healing capacities, they were consulted by women, married and unmarried, for enhancing their fertility. Writing in the early decades of the eighteenth century, Bartholomäus Ziegenbalg, the Lutheran missionary describes the ways in which *Komarams* functioned in South Indian coastal regions: in this case, at Tranquebar on the western coast.[3] Rich Freeman explains that similar oracles and oracle dances, as part of healing and diagnosing, have historically been present in the Malabar region for nearly two millennia.[4]

Another mother goddesses, *Cheranellur Durga*, who was venerated as *Mahishasura Mardhini* from early medieval times onward, was understood to possess curative agency in all fertility-related problems.[5] Writing in the sixteenth century, Duarte Barbosa describes how the *Chakliya*, a lower-caste group, worshipped another mother goddess, *Mariyamma* ("mother of diseases").[6] Believed to have been another form of *Kāli*, the quintessential mother goddess of wrath and anguish, women directed their prayers to her to beget healthy children.[7] Similarly, the traditions of fertility worship were found in the ritual complex of *Cheerma Bhagavati* at Payyannur *Kavu* (a sacred grove) as well, and her unhappiness and wrath were supposed to cause women to have premature deliveries and babies with deformities. These problems are believed to have been contained by the *Komaramas* who visited the houses in the locality, their sacred authority supposedly derived from the deity responsible for the cure. Religious rituals associated with mother goddess cults were considered to be preventive measures against contagious health disorders and epidemics. One such ritual, *mutiyettu*, alleviated the violent and divine whims of *Puthenkavu Bhagavati*. Similar rituals were performed to appease *Cheranellur Bhagavati* who was considered the incarnation of *Mahishasura Mardhini*, the most ferocious image of Durga, from as early as the fourteenth century, for various physical and mental diseases.[8] Her image, across India, had always been associated with motherhood, sex, and eroticism from the Gupta period onward. Social historians also point out that in Kerala and Bengal, *Mahishasura Mardhini* represents the annihilation of Buddhist practices and the appropriation of local fertility rituals.

Mother goddesses were thus perceived to possess the agency to enhance fertility enhancement and were worshiped by women through menstruation rituals such as the *Bharanippattu* (cock festival) at Kodungallur, which is known for its explicit sexual connotations. *Kodungallur Bhagavati*, another incarnation of Kāli was believed to inflict smallpox or chicken pox upon pregnant women. Being the most popular of the Bhagavati cults, she could ensure two kinds of fertility: earth fertility and female fertility. Sarah Caldwell describes her as "hot, full of rage, sexually dangerous, but she is also a loving mother whose blessing ensures prosperity and fertility. She is a mother, a virgin, and a warrior."[9] Even in the present day, pregnant women and potential mothers make offerings and special votive pilgrimages to please this goddess, who could otherwise transmogrify into the more aggressive form of *Bhadrakaali* to cause disdain, destruction, and barrenness.

The powerful tradition of sacred female cults induced in the common people a strong association with *yakshis,* immortal tree-bound sprites who, along with a host of other supernatural beings, also protected the local environment.[10] Forests have also been identified as the sources of powerful potions for use in sorcery and black magic from ancient times, and felling sacred trees is believed to have led *yakshis* to force pregnant women to feed on their own fetuses or kill their babies. Destroying vegetation was considered to disturb the spirits of the dead who resided in trees and could unleash the havoc of disease and infertility. The "Account of

Malabar," composed in the early eighteenth century, described the *yakshis* appearing in this manner:

> In the evenings under the form of fair young women; and these we call whorish devils, calling men with an audible voice, and as many as suffer themselves to be mastered by the voice of lust, and hearken to their voice, they murder them upon the spot; but they that fear them they assault and enter unto them, and they become *demoniacks* or possessed, and run about naked and mad to disturb the neighbourhood, eating grass and raw flesh.[11]

Apart from *yakshis*, people of medieval Malabar believed that the villages were surrounded by evil spirits such as *Chathan, Kuttichathan,* and *Ottamulachi* (a single-breasted spirit). The images of these spirits were worshipped at local temple complexes. Women in places such as Payyannur, north Malabar, sought divine blessings from local male deities such as *Peroorayyan*, who resided at *Perunmkavu* (great sacred grove), as described in the *Payyanurppattu*, a fourteenth-century song.[12]

It can be said that beliefs in these dreaded spirits created a parallel culture of body, medicine, and healing in the entire Malabar region. Native Christians, as well as Hindus and Muslims, regarded spirits as possessing the power to stimulate fertility. Barbosa writes:

> "When they are sick, that they send for such people to perform some ceremony whereby they hope to have their health restored; and at other times to help them to have children along with such other problems like discovering thefts all which are things repugnant in the authentic Christian religion."[13]

By the next century, the local Christian community had developed exclusive fertility cults, locating the cures performed by spirits at physical and psychological levels. They believed in the agencies of sprites to enhance fertility and ameliorate childbirth. These beliefs were described by Michael Geddes, in his late-seventeenth-century history of the Malabar church.[14]

Theyyam, a ritual dance form broadly associated with the lower-caste *Tiyya* community, was one among the various ritual practices that were closely associated with fertility cults. One such *theyyam*, the *Chamundi Theyyam* was performed to please *Chamundi*, the aggressive and strong mother goddess who, if enraged, could cause smallpox in pregnant women. *Theyyam* performers contained her occasional rage by walking on big heaps of burning charcoal, a practice that subsequently became the *curative ritual shield* against many such health issues. This form of dance became crystallized as a liturgical ritual during medieval times and became a living record of the lower-caste communities, their past, constituencies, and healing practices. A large number of fertility enhancers can be found in *theyyam* songs. Similarly, the expressions of fertility in *theyyam* rituals created a complex sense of non-textual, moral, and ritual hygiene concepts.[15]

Untouchable castes such as the *pulaya* also performed *theyyams* in their exclusive *kavus* as part of their education and socialization. The fertility rituals that they observed at the sacred *kavu*-temple complex at Meenkulam in the present-day Kannur district include feeding the sacred fish and Indian flap shell turtles in the temple ponds in order to cure the skin problems of newborn babies.[16] Other than the ritual performances associated with fertility issues, ritual foods at certain temples were also considered panaceas for all fertility-related matters. The turmeric *prasadam* of the Cheemeni Mundya temple at Kasargode is one such example. Some scholars suggest that medical or healing practices at such temples were of Buddhist origin and attest to the large-scale forceful Hindu appropriations of the medieval Buddhist *viharas* or *pallis* that were spread across in the region. Their arguments are based on the fact that traditional Hindu temples never offered treatments for any diseases inside their sacred complexes.[17] Temples such as Kurumba Bhagavati at Kodungallur or the Durga temple of Paruvasseri still bear testimony to their Buddhist pasts in their material structures and rituals.

Holy Bodies, Pilgrimage, and Fecundity Practices

Sacred geographies of all religions in South Asia created different kind of religious sociabilities and Healing, was a significant part of these spaces. Believers in all communities regarded pilgrimage as an important agency in enhancing the fertility potential of women. Although worship by native Malabar Christians did not parallel the variety and sophistication of saint worship in the Western Christian tradition during the medieval period, pregnant Christian women who sought fertility blessings undertook pilgrimages relevant to sacred Christian sites during this period. The Signora de Sauda (Our Lady of Health) Church at Tranquebar, six hundred kilometers from Calicut, was one such place and it evolved into a major site for receiving fertility blessings.[18] Female believers established communication with the divine figure at this sacred site through certain physical actions such as carrying with them clay pots, holy water, and written notes. When these objects of obeisance were fastened to their arms and necks, the power they created was a major hope for the fertility seekers.

Similarly, like any of their counterparts across Indian Ocean littorals and the subcontinental area, Muslims in the Malabar region worshiped their own holy figures for fertility-related issues. Sufi saints at places such as Idiyangara in the medieval port city of Calicut were major figures in this regard. Sheikh Masjid, a place where many Sufis are believed to have been buried, is one among many medieval Islamic sacred centers in Calicut. From the second half of the sixteenth century, this place became an enlarged sacred-healing complex for multiple illnesses, including all kinds of fertility issues.[19] A cultural-liturgical sphere developed around *nerchas* (celebrations of divine birth/death anniversaries) in which a curative space for fertility issues was also formed. The word *nercha* is derived from the Dravidian root *ner*, denoting multiple meanings including "truth," "agreement," and the act of taking a vow. Although influenced by some local ceremonial rituals, they were not merely internalizations, as Stephen Dale argues.[20] Most recent scholarship has allowed us to understand that the tradition of worshipping saints and seeking remedial measures from them had always been a part of Islamic communities worldwide. People, cutting across sociocultural differences, thronged at the sacred places of holy saints, during *nerchas*, singing hagiographical songs of which the earliest known is *Muhiyuddeenmala*, written in 1607 by Qazi Muhammed, the then religious authority of Calicut.[21]

The belief in martyrs and holy men as fertility healers with their *karamath* (miracles) goes back to the very early history of Islam in Arabia and in India. With the spread of *Malas*, performances of panegyric texts, celebrations of holy personalities, and *nerchas* garnered widespread social visibility and continuation among the

Muslim community. Subsequently, *nerchas* such as Kondotti *nercha* or "urs" (meaning death anniversary) remained blissful places for fertility seekers. Who here fertility seekers attributed their cure and energy to the votive "oil" and "cannons" offered at the *nercha* complex.[22] The curative guns that believers carried are still considered an ailment-cum-fertility shield, since the oil used in the guns is believed to have special medicinal properties and powers for fertility. Such beliefs and practices represent a continuation of the pattern that all Islamic communities have followed since the Abbasid period. The emergence of a new sociopolitical consciousness that took shape during British colonialism ultimately changed the pattern of the *nerchas* that earlier functioned as a space for communion and healing complex.

It is generally believed that the life of holy saints does not end with their physical disappearance from the worldly space, and their death simply means a new state of sacredness with more spiritual power for healing and miracles. As a result, Muslim *ulema* and scribal elites in Malabar produced several *mala* panegyrics during the period under discussion. These texts presented a number of charismatic sacred men and women from different Sufi traditions as healers and community heroes. These holy figures were believed to have special powers in the realms of pregnancy, childbirth, and fertility. The efficacy of *Sheikh Rifayee*, a prominent medieval Sufi who became quite popular among Muslims in the eighteenth century. In his interventions to promote fertility have been described in *Rifayee mala*, a *mala* panegyric written in 1781.[23] The suggestive effects of this saintly cult is believed to have had some influence on enhancing female fertility, male potency, children's survival chances at birth. Together, Sheikh Muhiyuddhin and Sheikh Rifayee were believed to have formed two *gawth* (axes) on which the whole "world turns." Thus, the presence of these saintly cults and related divine complexes in Malabar formed a large network of liturgical-curative shelters for women who were infertile, pregnant, and postnatal mothers. The continuity of this belief is seen in another panegyric text *Nafeesathmala*, written by Ponnani Nalakath Kunji Moideen Kutty in 1895.[24]

The fear of an augural evil eye or *ayn* was strongly associated with fertility psyche across South and West Asia, and Sufis' *karamath* (miraculous) power was sought for removing the peril.[25] An Arabic-Malayalam compilation of medieval healing, *Fee Shifau-n-Nasi: Ithu Orumichu Koottappetta Pazhaya Upakaram Tarjama Kitab* [hereafter *upakaram*], suggests that Muslims believed strongly in the principle of an evil eye, and considered this belief to be compatible with the textual tradition of Islam.[26] *Upakaram's* dedication of a large chapter to this issue suggests that, as in the case of most of their other health-related beliefs, Muslim healers in the region largely followed the Arab tradition of *ayn*. This text classified the *ayn* into *insiya* (humankind) and *djinniyah* (the *djin*), closely following the medieval Islamic classification, and believed that the evil influence of the *ayn* was transmitted through conscious stares, bodily touch, and verbal communications. Similarly, the *ayn* could cause sudden sickness, loss of temper and vigor, and grief in pregnant women and postnatal mothers.

The discursive textual tradition of Islam shows that Prophet Muhammed invoked Allah against the power of evil eye.[27] Ibn Qayyim Al-Jawziyya (1292–1350 AD), in his work on the Prophet's medicine, *Al-Tibb al-Nabawi*, argued at length about the logic behind the belief in the power of the evil eye and the counterpower of divinity.[28] His theory—that the *ayn* does not rest with the eye of a person, but rather with the *nafs* (spirit) of the beholder—resonated in some of the medieval Arabic medical texts that were in circulation in Malabar region during this period. Some of these texts are *Fawaid* by Imam Sharji' (1410–1488), *Al-Adkar Min Kalam Sayyad al-Abraar* by Imam Nawawi (1233–), and *Fath-ul-Malik* by Imam Dairabi. Muslims believed that persons with the evil eye could transmit *nafsnajsih* (impure spirit) to the body of pregnant women, instead of *nafs tahira* (pure spirit) of Sufis and healers. These impairments were removed by resorting to various ritual techniques such as reciting Suras *al-Mu'awwadhateyn* (*Sura, al-Falaq*—the day break; *Sura An-nass*—mankind), two small chapters of the Qur'an. Similarly, application of kohl to blacken the baby's eyelids or dimple for protection against the evil eye became part of Muslim life in Malabar, and newborn babies were smeared with various herbs to avoid the bad effects of the evil eye. Apart from their medical importance, the texts mentioned above and many others that fall outside the scope of this study show the predominant influence of various Arabic texts from the very early centuries of Islam in the region. These texts had played a significant part in establishing Malabar as a major area in the transnational Hadrami cosmopolis.

Similarly, Hindu women were to outwit the evil eye by carefully employing many hygiene rituals sanctioned by religion, both textual and non-textual. Bartholomaus Ziengenbalg describes that "a tender infant and newly delivered mother are particularly liable to the fascination of *evil eyes*, to the malign conjunctions of the planets, the influence of unlucky days, and many other dangers, each more perilous than [the] other."[29] Therefore, Hindu women thronged to *Bhagavati kavus* for partaking in *manthras* (chanting) to escape from evil-eye-related infertility issues. They also protected agricultural fertility by keeping enchanted objects or *molikas*, as described by the Dutch priest Jacobus Canter Visscher.[30] The person considered to be evil-eyed was socially and morally isolated and was avoided generally as he or she was considered an empty soul.

Likewise, in the realm of fertility, the causative agency of the *djin*, the celestial counterpart of a human being, was a major concern for Muslims. Women, pregnant or otherwise, were vulnerable in the presence of *Kafirdjins* who could cause infertility and pregnancy tensions, as well as destroying the female biological cycle. Possession by *djins* was believed to have caused substantial fertility problems such as miscarriage, which could be remedied by reciting panegyrics that invoked Sufis such as Sheikh Muhiyuddheen and Sheikh Rifayee. Noncommittal about a complete and successful healing through this method, *Upakaram*, the Arabic-Malayalam compilation of medieval and early modern Muslim healings, suggests that women who are infertile and less capable of childbearing should also consume certain sacred medicines or substances for enhanced fertility.

Several ingenious medicinal treatments point to a strong presence of *Tibb-un-Nabi* (prophetic medicine) in the realm of infertility in the post-high caliphate period in the Islamicate Indian Ocean. Similarly, in Malabar, fertility healing of *Tibb-un-Nabi* in which both medicine and divine alphanumeric symbols used, created a strong and parallel fertility Psyche. These sacred medicines consisted of Arabic letters and numerals mixed up in pictographic medicinal tables written in specially made Chinese ceramic bowls that were to be consumed after mixing with *zam-zam* (sacred water from Mecca). This practice, invariably called *Pinjanamezhuthu* (writing in ceramic) or more informally *ezhutheel kudikkal* (drinking the written), still exists across the Islamicate South Asia, including Malabar. The letters recommended were dependent on the character of the *djins* who were classified as "good" or "bad" in accordance with their spiritual status and behavior. In the sixteenth century, the Muslim scholar Sheikh Sainuddheen, in *Fath-ul-Muin*, recommended animal sacrifices in order to get away from the bad effects of a *djin*.[31] Interestingly, the beliefs in the causative capacities of a *djin* in the realm of body, diseases, possession, skin issues, and infertility continue to exist significantly among normal believers and *ulema*. Interestingly, a significant number of *ulema* from *salafis*, self proclaimed reformists believed in the curative and possessional qualities of *djins*.

Available sources suggest the Muslims on the Malabar coast were part of the larger Arab-Islamicate textual world, and were not really influenced by the *Unani* medical practices of North India where this form of Greco-Islamic medicine experienced a strongly institutionalized presence from the sixteenth century onward.[32] As a functional community, Muslims established themselves in Malabar in the twelfth century during the high caliphate in Bagdad where intense debates between Islamic theologians and medical scholars around the religious validity of *Unani* and Greco-Roman medicine took place. These debates created tensions around the whole knowledge system of *Unani*, which some section of the *ulema* considered taboo or un-Islamic. Malabar *ulema*'s disinterest in the humor-based Indo-Islamic medicine can be located in this context, as their ideas of healing derived from the Indian Ocean-based Arabic cosmopolis, not from the equally dominant "Persian cosmopolis" in which north India was part of. In Malabar, the urban *ulema* were more comfortable in their knowledge of prophetic medicine while the Mappila community in the towns was resorting to both prophetic medicine and indigenous medical practices, as attested in the subsequent vernacular textual tradition and practices.

Muslim *ulema* recommended medical herbs, including that of aphrodisiacs, when fertility issues such as male and female impotency arose and advised the community not to consider the use fertility medicines taboo.[33] Although methods involving aphrodisiacs were employed, magico-ritual healings were equally significant. *Upakaram* suggests that the belief in de-fertilization through *sihr* (blowing upon knots) was very strong in the Malabar Muslim community. This practice was a common form of witchcraft in the Arabian Peninsula, where women used to tie knots in the witch-cord and utter curses before they blew them off toward targets, real or imaginary. This, a witch-practice referred in the Qur'an, was countered by specific ritual techniques such as hanging of amulets and charms on the body, writing verses from the Qur'an, taking vows, and visiting the tombs of saints.[34] *Upakaram* elaborates the techniques of *sihr* healers who mastered the art of diagnosing and prognosing infertility issues.

Sihr healers were believed to possess the power of treating various sexual impairments such as impotency, sterility, and menstruation disorders through their techniques of synthesizing Qur'anic verses with Arabic alphanumerals. Healers were consulted for problems in pregnancy, childbirth, breast milk deficiency, problems during breast suckling, and diseases such as mastitis. Their mastery over Islamic scriptures, scribal works, and superior *dehavum puthayum* (body and cover) were believed to have enabled them becoming more effective.[35] Issues of *sihr* were treated with Qur'anic medicine, whereby specific chapters were used to contain the ailments. For instance, a certain arithmetic composition, the number of *hijra* months, and names of both pregnant lady and her mother, with the addition of the number 20, were used by *sihr* healers in their fertility engagements.[36] Muslim healers believed that if this calculation fell under the Qur'anic category of *lahv-ul-hayathu* (delightful life), the mother and child would live, and if the calculation fell under the imaginary category of *lahv-ul-mamath* (delightful death), both of them would die.[37] Similar techniques were employed in Malabar during the same period to prenatally-determine the sex of the child.

The culture of such magico-medical practices for causing and healing various illness and fertility problems was extensive in the region and practiced by all communities, as noticed by travelers such as Barbosa.[38] *Odiyan*, Kerala's unique shape-shifters, were believed to have possessed the power of de-fertilization with their mastery of black magic. Largely drawn from tribal communities such as the *Panans*, *Kadans*, and *Parayans*, they allegedly possessed the power of metamorphosing into animals and insects to harm fertility potential of women. They, who were also the low-caste masters of jungle pharmacopoeia, were seen as trafficking in the power of noxious supernatural beings and the substances of forests. They were suspected, accused, and at times executed for practicing sorcery. They were considered the antichrists of fertility and pregnancy, as they were believed to possess the power of removing pregnancy from young women by enticing and killing them. They were believed to have cultivated the power of killing pregnant women with *pillathailam* (child oil), which was said to be extracted from the body of an infant child or fetus of a young woman in her first pregnancy.[39] At the same time, Shafiite Muslim scholars in Malabar strongly prohibited abortion of any kind, unlike their medieval Hanafi counterparts who allowed it among slave women when they were impregnated by their Muslim owners.[40]

Snakes, Fertility, and Reproductive Hygiene

According to Balfour, writing in the late nineteenth century:

> Leprosy, ophthalmia, and childlessness are supposed by Hindus to be the punishment of men who in a former or present birth may have killed a serpent, and to be relieved of these the worship of the serpent is enjoined. The idea of their curative virtues is very old, and is still believed in India.[41]

Snakes, exotic and venomous, have been associated with sexual reproduction, fertility, and potency across South Asia from ancient times. Images of snakes inscribed on stones or their pictorial representations in temples have been worshipped across India. Abu'l-Fazl, the Mughal court chronicler, talks about ophite worship across north India during the sixteenth century. Snakes were considered an emblem of immortality, youthfulness, and fertility. The very stories of mythical formation of Kerala and the god Parasurama are associated with the snake god Nagaraja, who ensured abundance and perpetual land fertility in the region.

Subsequently, venomous snakes such as the *moorkhan* (cobra) and *mandali* (viper) inspired folk medicinal practitioners to connect them with the idea of sexual virility in the region. Barbosa talks about these two as being the most venomous found in the region in the sixteenth century. Hindus, across caste groups, adopted specific rituals and constructed temples for *moorkhan*, which was believed to have the power of affecting, enhancing, or reducing female fertility. In the medieval port-kingdoms, such as Calicut, the Samudris, the rulers of Malabar, took special care of this sacred snake which is known for its fecundity power. It is reported that when the king of Calicut "learns where the nest of any of these brutal animals is, he has made over it a little house on account of the water. If any person should kill one of these animals, they would immediately be put to death."[42]

Highly poisonous snakes such as *vellikkettan, iruthalakkuzhali, rudhiramandali*, and *karuvela* were venerated as symbols of life, female fertility, and death. From the twelfth century itself, rituals such as *Sarppam Thullal* (snake dance) are mentioned as part of the snake worship. These rituals expressed the associations of snakes with fertility and sexual energy. The extensive presence of various snakes in medieval Malabar also led to the growth of *vishaharis* (poison destroyers) who specialized in poison healing. *Vishaharis* found their way into the royal courts and rich men's houses where the constant fear of *vishamtheendal* (poisoning) was prevalent. *Pishari Kavu,* a medieval *kavu* belonging to the goddess *Badrakaali* situated in the medieval port town of Panthalayani near Calicut city has been consulted for protection from snakes. This sacred place, which developed into a major temple in the medieval period, obtained its unique name from its original name *Vishahari Kavu* (sacred grove for poison destruction). In his mid-nineteenth-century account, Edward Balfour says that death from snakebite was common across South Asia as people were not in a position to avoid snakes in their everyday lives. He notes that in the year 1861 alone British India witnessed 18,670 deaths from snakebites.[43]

The prominent presence of snakes in the fertility complex of Malabar gave rise to a large number of fertility proverbs that reflect the strong reverence for snakes in the fertility psyche of the region over a long period. For instance, the proverb *pampinte ina cheral pole* (like snakes make love) portrays deep and sensual intimacy between two people, while *anali petta pole* (like Russell's vipers' reproduction) refers to the healthy and excessive fertility rates of women in the region. The Russell's viper, being one of the viviparous snake varieties that gives birth to live young rather than eggs, has been considered a major symbol of fertility across the Indian Ocean littorals. The viviparous nature of this snake has always been a matter of human curiosity since very ancient times. In the Indonesian archipelago, the Russell's viper is also considered the most important image of agricultural fertility.[44]

Other proverbs, such as *alamuttiyal cherayum kadikkum* (even a rat, a snake bites as a last resort), *novichuvittamoorkhan* (like a harassed cobra), *neerkkolikum visham* (even a water snake carries poison), *keeriyum morkhanum pole* (like cobra and mongoose), *chanbhoomikku muzham pampu* (a yard of snake for a feet of land), *veliyilulla pampine tholilittu* (like shouldering snake lying on the fence), *theyyane thaccha pole* (like the beaten *teyyan*, an imaginary snake), *aleriyal pampu chavilla* (too many people seldom kill a snake), convince us of the strong sense of fear, reverence, and complicated emotions around snakes, poisonous or nonpoisonous. Several of these proverbs also demonstrate the place of snakes as major symbols of fecundity in the region. The spread of these proverbs across the communities even today shows that certain healing rituals were not community specific.

In Malabar, inextricably linked with fertility beliefs and rites were the complicated hygiene rules that can be categorized as social, moral, sexual, and ritual hygiene. Barbosa details the ways in which communities enforced caste- or religion-specific hygiene norms and sanctions for sexual intercourse and reproduction. Aberration from those rules invoked social/caste excommunication. The community of Brahmins ensured reproductive hygiene through a sanction that allowed only the eldest son in the family to marry while his brothers were allowed to practice *sambandham* (consortage) relationships with Sudra women from the Nair caste.[45] The children from these absentee husbands were not entitled to the right of inheritance and Brahmin fatherhood. In the sixteenth century, some of these sanctions were considered to be "strange customs" by contemporary Muslim scholars such as Sheikh Sainuddin Makhdhum.[46] However, as the spiritual authority of the Muslims and part of the Shafiite hadrami scholarly network in the Islamicate Indian Ocean littoral, Makhdhum himself had earlier presented a theological boundary for Muslim men and women in terms of their sexual, sensual, and bodily relations in the chapter titled *munakahat* in his text *Fath-ul-Muin*.[47] In *Fath-ul-Muin* we find ideas of early Islamic scholar al-Ghazali who describes piety, beauty, and fertility as the desirable qualities of a good wife. Also found in this text, in the context of marriage and female body, are references to the works of prominent medieval Shafiite scholars, such as Imam Nawawi, Imam Subuki, and Ibn Hajar. Hajar in his text *Sharah-ul-Irshad* prefers a highly fertile woman to an intelligent one in Muslim nuptial relationships.

The accounts of Barbosa and later travelers show the strict "physical" and "ritual" hygiene clauses that were maintained by higher caste groups in the area of reproductive intercourse. Sexual intimacy between higher-caste women and lower-caste men invited retribution and penal action, while it was permitted between higher-caste men and lower-caste women—provided the men followed certain purifying rituals after copulation. If a woman from a higher status was caught red-handed while committing prohibited sexual acts, they were forced to abandon their respective castes and leave the place, a practice that continued to exist until the nineteenth century. However, Brahmins were exempted from this rule as there was no question of casting them out since they were the main authority of defining and explaining sexual boundaries in accordance with their personal and communitarian interests.[48] Thus, *anuloma* intercourse (a man having sex with a woman of an inferior caste) in Malabar was related to a masculine fertility/reproductive psyche which accepted the children of upper-caste men and lower-caste women. However, the progeny of the upper-caste women with lower-caste men were believed to cause polluted conception and pregnancy and the result of such miscegenation was believed to destabilize the social structure itself.

Normally, upper-caste women were not permitted to be seen in public and in their everyday social life were shielded from external contacts. Nur Yalmon analyzes this as a deliberate strategy to preserve caste hygiene, which remained a significant condition for fertility purity.[49] "If a women has sexual contact with [a] lower caste male not only she and her future offspring but her caste could be polluted."[50] Barbosa reported that, in reaction, the lower-caste groups would try as hard as they could to touch upper-caste women by going to their houses at night: "And if they touch any woman, even if not witnessed, she, the Nair woman herself, was to publicise it immediately by crying out, and leaves her house without choosing to enter it again to damage her lineage."[51] These social hygiene punishments reflect the extreme level of caste protectionism and masculine sociability of the period. To a large extent, such sexual hygiene and fertility customs perpetuated the social dominance of the affluent in the larger cultural-liturgical establishments that ruled medieval social life in Malabar.

Fertility-related hygiene observations were largely entrenched in the territorially based social hierarchy, creating a social structure that decided hygiene enclosures from which many communities were excluded. Therefore, any violation of the sexual-moral hygiene norms, which were based on ritual ranking, was considered very seriously and often enforced with severe public punishments. Among the matrilineal communities, fertility rituals started off with *talikettu*, a significant puberty ritual during the period discussed. This ritual, which commenced at the onset of menstruation, ensured women's cultural initiation into the larger social complex and equipped her for marriage and further reproductive rituals. A woman who had not undergone this initiation ceremony was considered to be "polluting, a witch and dangerous for a man to marry because her sexual potency would be out of his control."[52] Rigorous hygiene and pollution rules were laid out across caste denominations during fertility cycles like menstruation, when women were considered particularly polluting.

Rules applying to menstrual periods, childbirth, and pregnancy were also meant to protect the fertility quotient of women during this time. One such rule says that "a newly confined woman has to stand at a distance of eighteen feet and a menstruating woman at twelve feet; hence the necessity in all respectable houses for special buildings set apart for special use by the women."[53] Thus, biological functions such as puberty, the menstrual cycle, and pregnancy kept women ritually secluded, thus acting to maintain a multitude of hygiene beliefs in the entire household. Parts of the fertility rituals were the specific rules designed to correct women's behavior, which were considered essential to bringing up disciplined children and society, at large. Girls who were undergoing puberty were perceived as vulnerable to the extra human powers that continue to violate the social hygiene of the dominant castes.[54] Thus, I would argue that in the Malabar region, the concepts around sexual pollution and sexual purity worked at two levels—the instrumental and the expressive. At the instrumental level, conspicuous actions were employed to monitor and control people's social and religious actions, while at the expressive level, belief and liturgical systems reinforced various hierarchical pressures.

Relations between Hindu reproductive customs and ritual hygiene were reflected in Barbosa's description of various forms of baths in medieval Malabar. Elaborate bathing rituals at various social and community initiation programs were conducted at puberty, marriage, and childbirth—apart from the regular daily bath—and the abundant spread of secular and sacred water bodies in Malabar ensured the sacredness of birth-related rituals and sanctions.[55] Similarly, in the *Fath-hul-Muin*, Sainuddheen Makhdhum gives elaborate rules and regulations for women after childbirth.[56] Taking a bath after the *nifas* (pregnancy blood) was mandatory under Islamic jurisprudence. Makhdhum stressed that the female body discharges multiple liquids during the process of delivery and purifying these was considered to ensure the health of the reproductive cycle as well. Vigorous application of sexual hygiene principles and fertility rituals on various occasions facilitated physical and social control over the female body. This text consider female body as more susceptible to sexual and reproductive pollution.

Apart from the normative hygiene rituals, existing literature also suggest that Hindu upper-caste women believed in a close association between high fertility rate and well-kept body. Medieval poets expressed their appreciation for high-caste women who kept their feminine side perfectly.[57] Many references are made to the ways in which they carried their embellished body in order to increase physical health and beauty.[58] Their health and beauty demands were met in the towns across port-kingdoms where people shopped for *mutthu* (pearl), *manikyam* (emerald), *rathnam* (precious stones), and *vajram* (diamond), and fertility-enhancing medicines.[59] They also realized the significance of smell in order to maintain a divine and hygienic environment where the female body could be spiritually cleansed. For the incineration of fragrance, they used *Ashtagandha* fragrances that created *dhoopa*, through which they tried to evoke the sacred bodies of the divine beings. This part of the rituals are mentioned in a twelfth-century text that describes the "sun worshipers" of medieval Kerala who took up "their censers and burn incenses in honour of fertility goddesses."[60] A combination of herbs, incenses, and fragrances were used to awaken these deities and appease the divine body of fertility saints.

Midwifery and Fertility

Fertility concepts, rituals, and cults on the south-west coast of Malabar were closely connected with midwifery traditions that became significant aspects of medieval village therapies and folk etiology.[61] Midwives or *petti* from specific lower castes were required to be present at the house of high-caste pregnant women from conception to delivery. Such caste-specific norms were believed to have ensured the continuity of high fertility, healthy reproductive mechanisms, and the well-being of the infant and the incumbent mother. Mostly, women from the lower-caste Malaya community performed midwifery, aided by their knowledge of fertility *mantras* and songs. Their songs, generally called *malayappattu*, were believed to have sacred powers to guarantee a healthy pregnancy and a smooth delivery. *Pettis* observed *maleyankettu*, a fertility dance ritual to avoid miscarriages, premature deliveries, and death of pregnant women. Bartholomaus Ziengenbalg says that "one of the principle reasons for which the European physicians are held [in] discredit in India, as far as regards their profession, is, that they administer their medicines without any *Mantra*."[62] *Mantra* recitation was a form of treatment, and with midwifery, existed among all sections of society during the period under consideration.[63] Christians also believed in the ritual role of the midwife. In the seventeenth century, Malabar Christians believed that *daia's* (midwives) ensured pure genealogy by preserving the ritual and religious purity of the Christian-born children from the "viciousness of Mahometans."[64] The *Fath-hul-Muin* suggests Muslims to have special fertility prayers to ensure children without satanic elements.⁶⁵ The emergence of community-based specializations in fertility and pregnancy-related issues also facilitated the community-specific fertility rituals. Men from the lower-caste *Vannan* community specialized in pediatrics and their women acted as obstetricians, while the *Malayan* community midwives remained the specialists in magico-ritualistic fertility treatments. People from the *Velan* caste, who were expert washer-men, kept the high-caste women from physical or ritual pollution throughout the course of their pregnancy with their cloths. The *Pulluvar* community, through their exclusive performance of the *pambinthullal/pulluvanthullal* (snake dance), granted fertility healings, especially when they were related to serpents.⁶⁶ When diseases such as convulsions and spasms affected pregnant women, they were to pray to sexually aggressive disease goddesses, such as *Aryakkara Bhagavati*,⁶⁷ by breaking into *thullal* (a dance performed while in a trance) in order to obtain instant cures.⁶⁸ Pushed into the social and geographical periphery through various social and cultural processes, communities such as *Pulaya, Malayan, Melan, Vannan, Pulluvan, Kuravan, Valluvan*, and so on developed their own tradition of sacred geographies, health practices, and fertility behaviors. These were significant in the given social milieu, despite the fact that Varthema, another traveler in the sixteenth century, considered most of these indigenous health practices to be "devilish."⁶⁹ Since their entry into the centers of humor-based medical learning and a larger social space that was believed to have polluted the entire social order of the medieval period, lower-caste communities produced their own therapeutics and reproductive healings through networks of small sacred spaces.

Similarly, the importance given to fertility, pregnancy, and childbirth produced strong traditions of *balavaidyam* (child care) in all communities during this time.⁷⁰ Child care became a specialized area of treatment toward the end of the period under discussion, when, as attested in the *Muhiyuddheenmala*, both premature deliveries and deaths at births were very high.⁷¹ Although the treatment of children became increasingly text-based in the upper-caste medical traditions, new specialist communities emerged at the lower level. Among the non-text-based specialist communities it was the *vannan* community that specialized in children's health.⁷² They were mostly called upon for the treatment of bronchitis and epilepsy and they treated the patients with many articles including medicinal plants.[73] The unique characteristic of this community was that they induced the power of *mantras* into the treatments for speedy recovery in a social situation in which common people sold their children out of poverty and diseases.[74]

Muslim *ulema* issued specific rules for child care, including child maintenance, postnatal food, and health care. Sainuddheen Makhdhum gave instructions based on the Islamic textual premise and insisted that women fed the infant with *labha'a* (colostrums). He suggested that the pre-milk with significant protective substances were essential for the infant's immune system.[75] This text also presents a series of rules and regulations with regard to the various stages of child development, and treated children as free entities. Muslims also conducted ritual sacrifices as a part of postnatal care. They were advised to conduct *Aqeeqa* or animal sacrifice between the birth and puberty of a Muslim child.[76] The "Account of Malabar," one of the early Danish accounts of Malabar in the early eighteenth century, devotes a full chapter to the diet of children in the Brahmin community. This text gives a proper picture of how children should be treated in terms of healthy food from the age of 5 to 15 years.[77]

In short, the attempt made here to examine the fertility practices in the Malabar region during the sixteenth and the nineteenth centuries clearly indicates that body and healing were not just seen from physical and practical perspectives there existed paradigms of culturally constructed of toward fertility practices and rituals. As mentioned earlier, both exclusiveness and crossovers in the fertility traditions of the Hindu communities show certain stringent sociocultural norms on one hand, and interdependency of the caste structure on the other. Although religious and caste communities attempted to create exclusive fertility zones, in actuality, fertility traditions remained interwoven and fluid, and most of the resulting healing complexes remained accessible to all communities. Likewise, the absence of foreign Muslims in the court of local rulers and the strife between Persianate Unani scholars and non-Persianate Islamic theologians in the medieval Islamic world perhaps reduced the emergence of an organized humor-based fertility healing method within the Muslim community as well. Malabar, as demonstrated above, was part of the larger Arabic textual network of Shafiite Islam, and Muslim *ulema* in the region resorted to *Tibb-un-Nabi*, which was developed and proliferated in the network of mosques and *dargahs*. Thus, these glimpses of fertility traditions, both textual and experiential, reveal a different dimension of the social and intellectual history of Malabar between 1500 and 1800: a long period that witnessed the emergence of new social bodies, bodily engagements, boundaries of emotions, and the emergence of new cultural-liturgical complexes.

Notes

1. Frederick M. Smith (2006), *The Self Possessed: Deity and Spirit Possession in South Asian Literature* (New York, NY: Columbia University Press), p. 547.
2. Sarah Caldwell (1998), "Bhagavati: Ball of Fire," in *Devi: Goddesses of India*, ed. John Stratton Hawley and Donna M. Wulff (Berkeley: University of California Press), p. 212; V. T. Indu Chudan (1969), *The Secret Chamber* (Trichur: Cochin Dewaswom Board), p. 56.
3. Bartholomäus Ziegenbalg (1682–1719), *Genealogy of the South Indian Gods: A Manual of the Mythology and Religion of the People of Southern India*, freely translated into English by Rev. G. J. Metzger (1869) (Madras: Higginbotham), p. 104.
4. J. R. Freeman (1999), "Gods, Groves and Culture of Nature in Kerala," *Modern Asian Studies*, 33 (2): 32–37; Rich Freeman (2008), "The Teyyam Tradition of Kerala," in *The Blackwell Companion to Hinduism*, ed. Gavin Flood (New Delhi: Blackwell), pp. 160–165.
5. Prof. Gopikkuttan (1996), *Koka Sandesham* (Malayalam translation) (Trichur: Current Books), line 92; Ralph W. Nicholas (1981), "The Goddess Sitala and Epidemic Small Pox in Bengal," *The Journal of Asian Studies*, 41: 21–44; Babagrahi Misra (1969), "'Sitala': The Small Pox Goddess of India," *Asian Folklore Studies*, 28 (2): 133–142.
6. Duarte Barbosa (2009 [1865]), *A Description of the Coast of Africa and Malabar in the Beginning of the Sixteenth Century*, trans. and ed. Henry E. J. Stanley (London: Hakluyt Society), p. 115.
7. K. Rathnamma, Compilation and Commentary (1997), *Ananthapuravarnanam* (Trivandrum: State Institution of Languages), line 143.
8. Gopikkuttan, *Koka Sandesham*, line 92.
9. Caldwell, "Bhagavati: Ball of Fire," p. 196.
10. William Dalrymple (2009), "The Dancer of Kannur," *Nine Lives: In Search of the Sacred in Modern India* (London: Bloomsbury).
11. J. T. Philips, trans. (1717), *An Account of the Religion, Manners, and Learning of the Malabar*, translated from the Dutch (London: Black Swan), pp. 84–85.

12. P. Antony, ed. (2000), *Payyannur Pattu*, Tuebingen University Library: Malayalam Manuscript Series, General ed., Scaria Zacharia (Kottayam: DC Books), pp. 9–10.
13. Duarte Barbosa (2009), *A Description of the Coast of Africa and Malabar in the Beginning of the Sixteenth Century*, trans. and ed. Henry E. J. Stanley (London: Hakluyt Society, 1865), p. 176.
14. Michael Geddes (1694), *The History of the Church of Malabar*…(London: Printed for Sam. Smith and Benj. Walford).
15. J. R. Freeman (1999), "Gods, Groves, and the Culture of Nature in Kerala," *Modern Asian Studies*, 33 (2): 257–302, 285.
16. *Records of the Zoological Survey of India* (2004), Vol. 102, Issues 1–2 (New Delhi: Zoological Survey of India).
17. P. K. Gopalakrishnan (2000), *Keralathinte Samskarika Charitram* (Trivandrum: State Institute of Languages), p. 255.
18. Niklas Thode Jensen (2005), "The Medical Skills of the Malabar Doctors in Tranquebar, India, as recorded by Surgeon T L F Folly, 1798," *Medical History*, 49 (4): 489–515, 503.
19. Ziyaud-Din A. Desai (1989), *A Topographical List of Arabic, Persian, and Urdu Inscriptions of South India* (New Delhi: Indian Council of Historical Research), p. 103.
20. Stephen Dale and Gangadhara Menon (1978), "'Nerccas': Saint Martyr Worship among the Muslims of Kerala," *Bulletin of the School of Oriental and African Studies*, 41 (3): 523–538, 525.
21. Qazi Muhammed (2000 [1607]), *Muhiyuddeen Mala* (Calicut: Thirurangadi Book Stall), lines 75–85.
22. Dale and Menon, "'Nerccas.'"
23. (2000 [1781]) *Rifayee Mala* (Calicut: Thirurangadi Book Stall), pp. 72–76.
24. For more details see, P. K. Yasser Arafath (2012), "History of Medicine and Hygiene in Medieval Kerala: 14–16 Centuries," Unpublished PhD thesis, University of Hyderabad, Hyderabad.
25. Ponnani Nalakath Kunji Moideen Kutty (2004 [1895]), *Nafeesath Mala* (Calicut: Thirurangadi Book Stall), p. 4; E. Hobsbawm and T. Ranger (1983), *The Invention of Tradition* (Cambridge: Cambridge University Press), p. 465.
26. Ahmad Bava Musliar, ed. and annotated (2001 [1885]), *Fee Shifau-n-nasi: Ithu Orumichu Koottappetta Pazhaya Upakaram Tarjama Kitab* ("This translated compilation contains remedies for people"), a compendium of medieval healings, edited and annotated in 1885 by and reprinted at C. H. Muhammed Koya and Sons, Thirurangadi, p. 81.
27. Imam al-Bukhari (810–72 AD), in his *Book of Medicine*, recorded that the Prophet one day saw a servant girl with a *sa'fa* (or suf'a, a black or brown mark or excoriation) on her face, and said: "Recite incantations for her, for the 'glance' is on her," translated by Muhammad Muhsin Khan (1997), *Sahih al-Bukhari* (Riyadh: Darussalam), p. 426.

28. Abd al-Ghani Abd al-Khaliq, ed. (1957), *Muhammad ibn Abi Bakr ibn Qayyim al-Jawziyya, al Tibb al-Nabawi* (Beirut: Dar al-Kutub al Ilmiyya), pp. 127–136.
29. Ziegenbalg, *Genealogy of the South Indian Gods*, p. 104.
30. Jacobus Canter Visscher (1862), *Letters from Malabar (1743) tr.: To Which Is Added an Account of Travancore, and Fra Bartolomeo's Travels in That Country* (Madras: Heber Drury), p. 145.
31. Sainuddheen Makhdhum (2012 [1575]), *Fath-hul-Muin* (Calicut: Poomkavanam), p. 304.
32. Seema Alavi (2008), *Islam and Healing: Loss and Recovery of an Indo-Muslim Medical Tradition 1600–1900* (Ranikhet: Permanent Black), pp. 18–28.
33. Makhdhum, *Fath-hul-Muin*, p. 495.
34. Aref Abu-Rabia (2005), "The Evil Eye and Cultural Beliefs among the Bedouin Tribes of the Negev, Middle East," *Folklore*, 116 (3): 241–254, 247.
35. *Upakaram*, p. 60.
36. Ibid., pp. 59–62.
37. Ibid., pp. 53, 60.
38. Barbosa, *A Description of the Coast of Africa and Malabar*, p. 142.
39. A. Sreedhara Menon (2008), *Cultural Heritage of Kerala* (Kottayam: DC Books), p. 105; F. Fawcett (1985), *Nayars of Malabar* (New Delhi: Asian Educational Services), p. 311.
40. Makhdhum, *Fath-hul-Muin*, p. 587.
41. Edward Balfour (1885), *The Cyclopædia of India and of Eastern and Southern Asia*, Vol. 3 (London: B. Quaritch), 3 vols., pp. 569–580.
42. Ludovico Di Varthema (1863 [1510]), *The Travels of Ludovico Di Varthema in Egypt, Syria, Arabia Desert and Arabia Felix, in Persia, India, and Ethiopia, AD 1503 to 1508*, trans. John Winter Jones (London: Hakulyt Society), pp. 173–174.
43. Balfour, *Cyclopædia*, p. 573.
44. D. G. Blackburn and J. R. Stewart (2011), "Viviparity and Placentration in Snakes," in *Reproductive Biology and Phylogeny of Snakes*, ed. R. D. Aldrich and D. M. Sever (New Hampshire: Science), pp. 119–181.
45. First Travancore Nair Act of 1913 recognized *Sambandham* as a legal marriage and allowed wife and children of a Nair, dying intestate, one half of his self-acquired property, says C. J. Fuller (1976), *The Nayars Today* (Cambridge: Cambridge University Press), p. 134; for a detailed discussion on the system of medieval Consortage, see J. Devika (2005), "The Traditional Nayar Marriage System," *Modern Asian Studies*, 39: 99–123.
46. Sainuddin Makhdhum (1999 [1583]), *Tuhfat-hul-Mujahideenfeen Ba-a-Si Akhbaril Burthukhaliyeen*, translated as "A Tribute to The Warriors with Information about Portuguese" (Malayalam translation), ed. C. Hamsa (Calicut: Al-Huda Book Stall), p. 41.
47. Ibid., *Tuhfat-hul-Mujahideenfee*, pp. 454–573.

48. P. Bhaskaranuni (2000), *Pathompatham Nootandile Keralam* (Trissur: Kerala Sahitya Akademi), p. 12.
49. Nur Yulman (1963), "On the Purity of Women in the Castes of Ceylon and Malabar," *Journal of the Royal Anthropological Institute*, 93: 25–58.
50. Karen Paige and Jeffery M. Paige (1981), *The Politics of Reproductive Ritual* (California: University of California), p. 24.
51. Barbosa, *A Description of the Coast of Africa and Malabar*, p. 143.
52. Susan Lipshitz (1978), *Tearing the Veil: Essays on Femininity* (London: Routledge & Kegan Paul), p. 48.
53. William Logan (1951 [1887]), *Malabar Manual* (New Delhi: Asian Educational Service), p. 118.
54. Uma Chakravarti (2004), "Conceptualizing Brahmanical Patriarchy in Early India; Gender, Caste, Class, and State," in *Readings in Indian Government and Politics, Class, Caste, Gender*, ed. Manoranjan Mohanty (New Delhi: Sage).
55. Ludovico Di Varthema, *Travels*, p. 149.
56. Makhdhum, *Tuhfat-hul-Mujahideenfee*, pp. 79–80.
57. Gopikkuttan, *Koka Sandesham*, pp. 59, 64, 89.
58. Antony, ed., *Payyannur Pattu*, p. 30.
59. Ibid., pp. 29–30. "A lucky venture! A lucky venture! Plenty of rubies, plenty of emeralds, many thanks you owe to God for bringing you to a country where there are such riches! Says Emnome de Deus (2009), *The Journal of the First Voyage of Vasco da Gama to India, 1497–1499*, trans. Glenn J. Ames (Leiden: Brill), p. 72; Barbosa, *A Description of the Coast of Africa and Malabar*, p. 31.
60. Rabbi Benjamin of Tudela (1840 [1167]), *The Itinerary of Rabbi Benjamin of Tudela*, trans. and ed. A. Asher (London: A. Asher), p. 140.
61. Antony, ed., *Payyannur Pattu*, line 17.
62. Ziengenbalg, *Genealogy of the South Indian Gods*, p. 103.
63. M. K. Vaidyar, ed. (1951 [1800]), *Mahasara*, A compilation of 33 medieval texts by an unknown person in the beginning of 1800, with a preface by M. K. Vaidyar (Madras: Government Oriental Manuscript Library), p. xvi.
64. Geddes, *The History of the Church of Malabar*, p. 196.
65. Makhdhum, *Fath-hul-Muin*, p. 495.
66. Edgar Thurston and K. Rangachari (1909), *Castes and Tribes of Southern India*, 7 vols. (Madras: Government Press), Vol. 4, p. 228.
67. Barbosa, *A Description of the Coast of Africa and Malabar*, p. 57; K. K. N. Kurup (1977), *Aryan and Dravidian Elements in Malabar Folklore: A Case Study of Ramavilliam Kalakam* (Kerala Historical Society: Distributors; Trivandrum: College Book House), p. 19.
68. Fra Paolino Bartolomeo (1776–1789), *Voyage to the East Indies; Observations Made during a Residence of Thirteen Years between 1776 and 1789 in District Little Frequented by the Europeans*, trans. William Johnston, (London: J. Cuthell), pp. 405–406.
69. Varthema, *Travels*, p. 167.

70. Sriman Nambothiri (1990), *Chikitsa Manjari* (trans-Malayalam) (Alapuzha: Vidyarambham), p. 442.
71. Qazi Muhammed, *Muhiyuddeen Mala*, line 112.
72. Koramangalam Narayanan Nambhutiri (2007), "Natuvaidyahile Vannan Paramparyam," in *Natarivukal: Natuvaidym*, ed. Hafeel (Kottayam: DC Books), pp. 80–89.
73. Ibid., p. 82.
74. Barbosa, *A Description of the Coast of Africa and Malabar*, p. 180.
75. Makhdhum, *Fath-hul-Muin*, p. 573.
76. Ibid., p. 299.
77. Philips, trans. *An Account of the Religion Manners, and Learning of the Malabar*, p. 146.

4

The Circulation of Medical Knowledge through Tamil Manuscripts in Early Modern Paris, Halle, Copenhagen, and London

S. Jeyaseela Stephen

The Tamil littoral occupies a very small space in the vast expanse of the Indian Ocean. The northern segment of the Coromandel Coast was famous for the production of cotton and the export of textiles. The southern portion, sometimes called the Pearl Fishery Coast, was well known for its pearl banks and the export of chanks and pearls through Asian maritime trading networks. The European trading in the region and the religious activities of the missionaries in the early modern age resulted in the exchange of scientific knowledge. The medical encounter, in particular, and Euro-Tamil relations, more broadly, witnessed extraordinary and unprecedented innovations between 1700 and 1857.

Several scholars have significantly contributed toward an understanding of medicine in the age of commerce and empire. Harold J. Cook's study of the involvement of the Vereenigde Oost-indische Compagnie (VOC) with the scientific revolution in the Low Countries demonstrated the role of several Dutch physicians who traveled to the Eastern colonies, transferring natural and medical knowledge through written treatises and preserved plant and animal specimens, and carrying medicines to be sold in pharmacies and living plants to be planted in botanical gardens.[1]

David Arnold has studied the European reactions to diseases of warm and hot climates, underlining the entanglement of tropical medicine with the political and social aspects of empire. As he notes, substantial colonial literature begins to emerge after 1750: a time during which the broader image of the tropics in Europe was shifting from utopian imaginings toward concepts of tropical nature as "malevolent" or even hellish.[2] Pratik Chakrabarti explored how medicine was transferred in the eighteenth-century British Empire, aligning the trajectories of intellectual and material wealth. He argues that medicine acquired a new materialism as well as new materials in the world of global commerce and warfare.[3]

While these authors provide useful frameworks for thinking about the transmission of medical knowledge, in general, to European practitioners, none focus exclusively on the Siddha system of medicine. In this chapter, I consider the development, the scope, and the character of European studies of Tamil medical texts. I also examine how several new medical treatises came to be composed both by European and by Tamil physicians in the age of European expansion. This chapter focuses also on the spread of Tamil medical knowledge through translation of medical texts from Tamil into French, English, Danish, and German. It traces how Europeans were motivated to purchase and transport the palm-leaf medical manuscripts written in Tamil, which ultimately found their way to various repositories in Europe.

As in other parts of India, Brahmins in the Tamil country tended to make use of Ayurvedic medical texts composed in Sanskrit in the ancient period. Other elite medical texts used in the region employed the Telugu language. However, Siddha medical texts written in Tamil in the medieval and early modern periods remained widely popular among all other castes when the Europeans arrived on the coast. During their apostolic work, the European missionaries noticed the significance of the Siddha healing practices of the local medical practitioners of several castes that formed a part of their culture. Some English East India Company (EIC) physicians and surgeons had high praise for Siddha medicine. This judgment was based on their experience of using remedies drawn from Tamil pharmacopoeia in the European hospitals in the settlements in India as well as their integration into the practice of medicine in Europe. European missionaries and some members of the EIC ventured into the study of the Tamil language and some made sincere and serious efforts to understand Siddha medicine and the Tamil cultural heritage. Examining how the ideas of the Siddha system of medicine that reached Europe gained acceptance, modification, and rejection not only sheds light on medical exchanges between Hindus and European missionaries, but also provides a cultural perspective on the spread of Christianity in the Tamil country.

Jesuits and the Study of Tamil Medical Texts

Jesuit missionaries preached the Gospel in the Tamil coast, learning Tamil from the time of Roberto Nobili, who worked at the Madurai mission in the early years of the seventeenth century. By the late eighteenth century, the Jesuits were focusing their conversion efforts mainly on people of the lower strata of society. They evinced interest in establishing dispensaries and set up pharmacies, prepared medicine, and supplied it free of charge to residents of the so-called black towns of the European settlements. Therefore they showed an abiding interest in the study of Tamil medical texts. One medical text that attracted the attention of the Jesuit missionaries was the *Sillarai Kovai* ("Miscellaneous Collection"), a text that dealt with the subject of toxicology, focusing, in particular, on antidotes for poisons, and remedies for bites of various types of poisonous snakes and insects as well as dogs.[4]

Gaston Laurent Coeurdoux, who was in Pondicherry in 1740, carried out apostolic work in the French settlement on the Coromandel Coast,[5] becoming fluent in Tamil. He developed an interest in Tamil toxicology, observing that the local Siddha medicines prepared and administered to the patients were more effective than those imported from France, and as a result of his interest came across the Siddha medical text *Sillarai Kovai*. Based on this text or on his observations of cures being practiced, he sent what he described as an excellent remedy against snakebite to Souciet, the Jesuit Superior General in Rome in 1738.[6]

Another Jesuit, Friar Jean-Baptiste du Choiseul, arrived in Pondicherry in 1740.[7] He became the apothecary in the pharmacy of Pondicherry in 1742 after the death of Friar Alexis Mazeret who had previously occupied the post.[8] He developed an interest in Siddha medicine, which he studied in some detail, especially its teachings on the subject of poisonous bites. He also found *Sillarai Kovai* very useful and he culled information from the manuscript, writing a treatise in French, the title of which translates as "A New, Sure, Short, and Easy Method of Treating Persons Affected by Rabies." Later he sent the same treatise in the ship that sailed from Pondicherry to France. It was published as a book in 1756 in Paris.[9] Since it contained very useful data, Louis Pasteur (1822–1865), the French biologist and chemist who became interested in this particular field, consulted it. The book is preserved in the *Archives de la Province de France de la Compagnie de Jesus*, Vanves, Paris.[10] We know that Pasteur was highly successful in his discovery of the vaccine for rabies. He also laid the foundations of experimental methods in microbiology, highlighted the role of microbes in the propagation of disease, and invented pasteurization. Thus Siddha medical knowledge transmitted to France found both direct and indirect applications. The dissemination of Siddha knowledge to France by the Jesuits on the Tamil coast should be regarded as a milestone in early modern medical history.

Protestant Missionaries and the Study of Tamil Medical Texts

Several Protestant missionaries also came to preach the Gospel on the Tamil coast, the first being Bartholomäus Ziegenbalg. He arrived at Tranquebar, the Danish settlement in the Tamil coast, on July 9, 1706.[11] He learnt Tamil and met a Tamil physician at Tranquebar for the first time on May 1, 1708. He requested the native doctor to permit him to glance at some Tamil palm-leaf medical manuscripts. Under the guidance of this physician, Ziegenbalg was allowed access to several texts on Siddha medicine.[12] He thereafter developed an interest in collecting medical manuscripts. As his understanding of Tamil improved, he also tried to learn more about medicine. He was fully convinced of the usefulness of the contents and significance of a palm-leaf medical manuscript titled *Waguda Tschuwadi* (*Vaagada Chuvadi* in Tamil, or Manual of Medical Treatment) through contact with the Tamil physician.[13] As he found extreme difficulty in reading and understanding the verses of the medieval medical text, he sought the assistance of a native to comprehend the contents. He was very successful in his attempt. According to Ziegenbalg, the *Vaagada Chuvadi* contained 80 sections of 120 chapters in 6 volumes. The first volume, *Sutra Sthanam*, in 30 chapters, dealt with the etiology of diseases. The second volume, *Sarira Sthanam*, in 16 chapters, dealt with physiology and anatomy. The third volume, *Kaarana Sthanam*, in six chapters, dealt with some aspects of pathology. The fourth volume, *Indira Sthanam*, in 16 chapters, dealt with diagnosis and prognosis. The fifth volume, *Kalpa Sthanam*, in 11 chapters, dealt with toxicology and antidotes. The sixth volume, *Sikitchai Sthanam*, in 41 chapters, dealt with therapeutics. An appendix to this large text itself consisted of four chapters. Zieganbalg was highly impressed by the rich store of medical ideas contained in the manuscript, so he purchased and sent the *Vaagada Chuvadi* manuscript in a ship that sailed from Tranquebar to Denmark in 1709, which found its way finally to Halle, the mission headquarters in Germany, in 1710.[14] The verses of this text of Agasthiyar *Vaagadam* were written in *viruttam*, an improvisational form used for devotional songs. It gives information regarding medicinal salts, acids, and waxes.

Ziegenbalg opined that the Tamils were extremely advanced in their medical studies. In his opinion, the medical works he consulted contained information on vital matters that were really of practicable use. The basis of these Siddha medical works was *tridosha siddhantam*, the teaching that diseases were related to *vaatham* (wind, connected to the gastric system), *pitham* (bile), and *kapam* (phlegm, connected to the respiratory system). A person's susceptibilities to imbalances of these vital fluids developed while in the mother's womb. All diseases could be grouped under these three basic categories. Correspondingly, they were detectable through three different pulse systems. All these details were written in the medical text *Vaagada Chuvadi*. The three types of pulses could be felt at the wrist, neck, and feet knuckles. For women it could be felt when examined on the pulse in the left wrist. For men it was on the right wrist. The *vaatham* pulses beat like frogs, the *pitham* pulses beat like hens, and the *kapam* pulses beat like peacocks. If the *vaatham* pulse and the *pitham* pulse beat likewise, then the patient had sore throat, cough, running nose, perspiration, and heat in the body. Ziegenbalg stated that the urine test conducted in the Tamil country resembled that used in Europe, but he went on to describe its specific features in detail. The doctor first examined the patient to judge whether the condition was likely to be critical or not. The patient was asked to pass urine in a particular vessel. A certain type of oil (sesame oil) was taken up in a straw and then allowed to trickle down into the vessel. If the oil drops floated, then it was concluded that the patient's life could be saved. If the oil drops sank to the bottom of the vessel, then it was realized otherwise.[15]

Another important medical manuscript in Tamil that had attracted the attention of Ziegenbalg was *Udelkudu Wannam* (*Udalkuru Vannam* by Dakshinamurthy Siddhar), a text that dealt with questions of the nature of human life and with anatomy. It also dealt with the theory of functions of a human body as a vehicle attaining salvation and mentioned the power of the five sense organs. He therefore purchased it and sent it to Halle. According to Ziegenbalg, the anatomical teachings contained in the text were understood only by a few Tamil doctors. Despite the references to dissection in some ancient works such as the Susruta Samhita, contemporary practitioners of Ayurveda were comparatively ignorant of anatomy, given the modern aversion to the dissection of the human body.[16] The two manuscripts that Ziegenbalg sent back to Halle, *Waguda Tschuwadi* and *Udelkudu Wannam*, are preserved even today at the *Archiv der Franckesche Stiftungen*, at Halle.[17]

Johann Ernst Grundler, the missionary, came to Tranquebar on July 20, 1709.[18] He learnt Tamil and developed an interest in studying palm-leaf manuscripts on the subject of medicine. He shifted his residence from Tranquebar to Poraiyar, a village nearby, on February 20, 1710.[19] He spent an unbroken period of 11 years there, studying medical manuscripts. In order to know Tamil culture more intimately, he started eating and dressing like the Tamils, following the earlier custom of Jesuit missionaries. He was mainly interested in the diseases, diagnoses, and dosages found in Tamil medical texts. He noted carefully how the Tamil physicians diagnosed the diseases by pulse, phlegm, urine, tongue, and eye tests and how they prescribed proper medicine by identifying certain special characteristics of the diseases and relating them to three separate groups of people: men, women, and children. He desired to disseminate the rich medical knowledge he found in these texts, so he translated the palm-leaf medical text *Agastiyar Irandaayiram* from Tamil into German. He stated that as a translator he wished to be of service to European medical men. He wrote the treatise I refer to here as "The Tamil Physician" in German and he completed it in 1711. The text was sent by Grundler to Halle in Germany via Copenhagen by the ship that left Tranquebar in 1712. This manuscript, titled *Der malabarische medicus, welcher kurzen Berricht gibet, theils was diese Heyden in der medicin vor Principia haben; theils auf was Art und mit welchen Malabaren Zusammen getragen u.ubersetzet von Johann Ernst Grundler*, is preserved in the *Archiv der Franckesche Stiftungen*, Halle.[20] Grundler mentioned in this work that he had compiled and translated the details from the Tamil manuscript into German. He had also furnished details related to the circumstances under which medicines were given by the natives to cure diseases. In his preface he gave a short account, obtained from a Brahmin, explaining how studies in medicine were conducted and taught to pupils in Tamil schools.[21]

The manuscript "The Tamil Physician" has two parts. The first part contains ten themes: (1) arrival of the doctor and the arrangements to be made to attend the sick; (2) on pulse and its types; (3) how to recognize the main disease; (4) the diseases that emerge from the *tridosha siddhantam*: *vaatham, pitham,* and *kapam*; (5) the origin of those diseases; (6) the diets that should be observed in these diseases; (7) on fever; (8) on unconsciousness/fainting (syncope); (9) on the symptoms that indicate whether a patient will live or die; and (10) the manners of diseases that occur among the Hindus.[22]

The second part of the manuscript contains 12 themes: (1) how to purify and pulverize some plants to make medicines; (2) medicines by which fever can be cured; (3) medicines by which fainting can be cured; (4) medicines by which different poisonous diseases and ulcers can be cured; (5) head diseases; (6) eye diseases; (7) diseases of the ears, nose, and mouth; (8) diseases of the chest and stomach; (9) on diarrhea; (10) on venereal and several other urinary diseases; (11) on female diseases; and (12) on children's diseases.[23] Finally, we find notes to parts one and two appended, along with a dictionary of medical terminology.[24]

Grundler elaborated on the different kinds of medicines used by Tamils for the cure of various diseases. The prescription details contained the ingredients used to make medicines, such as *yellu* (*Sesamum orientale*), *vashambu* (*Acorus calamus*), *yerukkai*

(*Calotropis gigantea*), *nochi* (*Vitex negundo*), *agatthi* (*Sesbania grandiflora*), *kaatunelli* (*Phyllanthus emblica*), *inji* (*Zingiber officinale*), *murungai* (*Moringa pterygosperma*), *saaranay* (*Trianthema portulacastrum*), and such as like *induppu* and *malaiuppu*. The old usages of weights and measures were also given in the text.

On analysis and scrutiny of Grundler's text, I noticed that he had exactly followed the pattern and details of the Tamil manuscript *Agastiyar Irandaayiram*.[25] The Tamil work conforms precisely to Grundler's account: part one, dealing with the doctor and disease; and part two, dealing with medicines for various diseases. Grundler's work was used by those Germans who had evinced interest in the Siddha medical system. The doctors read, used, and taught the Tamil medical system in Copenhagen as early as 1713. Grundler subsequently wrote extensively on the Siddha system and its rules for the preparation of medicines, and his interest is also evident from letters to friends in Germany.[26] Thus, Siddha medical knowledge was communicated to a wider audience, while Grundler also passed on specific pieces of information from time to time.

The EIC Surgeons, Physicians, and the Study of Tamil Medical Texts

Theodor Ludvig Frederich Folly was the Danish surgeon at the military hospital in Tranquebar in 1777.[27] He was interested in the details of Siddha medicine and he studied the medical manuscripts through a local assistant. He developed a particular passion to study the medical texts of Agasthiyar. He composed a series of essays in Danish during 1795 and 1801, but these were not printed.[28]

There exists a manuscript, now preserved at the Rare Manuscripts Section in *Det Kongelige Bibliotek*, Kobenhavn.[29] There is also another manuscript at Kirurgisk Akademi, Kobenhavn.[30] Both are in Danish, titled *Anmærkning Om Mallebar Lægernes Kundskab i Chirurgien, 1798*. In these texts, Folly had translated the process of mercury sublimation from Agasthiyar's manuscript. He stated that the purpose for his painstaking work of translation was to help Danish and German doctors. It states that one *palam* (i.e., Tamil ounce, of which 13 makes up one English pound) or the weight of 20 *pagodas* of *singaram* (alum) and 2 *palams* or the weight of 20 *pagodas* of *rasam* (mercury) had to be taken and rubbed together for 4 *saamam* (12 hours) with the juice of *pasu manjal* (*Curcuma domestica*), *thulasi* leaves (*Ocimum sanctum*), *alangi* root (*Alangium salvifolium*), and lemon fruit. Then this mass had to be poured into a *kasi kuppi* (a glass flask of Benares) and wrapped up with old rags, on which the *kaliman* (literally "clay" in Tamil, but referring here to a sort of lime prepared out of small ground gravel stones mixed with jaggery, egg white, sour milk, oil, and butter) had to be mixed and rubbed. A stone or a piece of an old clay pot had to be put over the mouth of the flask, and wrapped with the rags. The flask had be placed in the sun to dry. Afterward, a fireplace in the ground had to be dug. A *saddi* (clay pot) had to be placed over it. Then sand had to be put underneath, in the fireplace, to a height of two fingers' width. The flask had to be placed into it, so that it stood upright. Then the flask had to be burnt with a *deepam* (lamp), at first for four *saamams* (12 hours), again with a *kamalam* (a metal plate that was used as an oil lamp with ten wicks) for four *saamams*, and then with firewood for four *sammams*; in total, the fire had to last for one and a half days. Afterward it had to be allowed to cool. Then the flask had to be broken, and then *rasakarpuram* (the mercury) would be found. According to Folly, this mercury was an excellent drug and used for the treatment of syphilis in Siddha medicine. He finishes by cautioning that what he had translated from the Tamil Siddha medical manuscript about the preparation of mercury by the Tamils might not be acceptable to the learned men of Europe.

Folly attempted, in October 1799, to collect medical manuscripts attributed Agasthiyar, Bogar, Karuvurar, Konganavar, Sattainathar, Theraiyar, Tirumular, Punnaikisar, and Pulasthiyar, the nine great Siddhars or founding saints of Siddha medicine. He said that these manuscripts contained descriptions of the preparation of medicines, both simple and composite. He stated that he was not successful in collecting the manuscripts because they were then not available from the physicians he had contacted. T. L. F. Folly also had extensive discussions with two Tamil doctors (father and son) who came from Madurai to Tranquebar to sell medicine. From them, he learnt the details of the preparation of Tamil herbal medicine. In return, Folly gave them one pound of purified mercury, which pleased the old man. Mercury was being imported from China by the European traders to the Tamil coast and had widespread use in both Indian and European medicine during the eighteenth century.[31]

Benjamin Heyne was appointed as a surgeon to the Moravian Mission in 1790 and reached Tranquebar in 1792. He was there for two years and then entered the services of the English EIC and was appointed first as a botanist and later as an assistant surgeon in Madras in 1799.[32] He was entrusted with the responsibility of suggesting and prescribing the bazaar medicines that were needed for the supply of the English Company's army.[33] This gave Heyne an occasion to study Siddha medicine in detail.

The English Company at Madras welcomed surgeons and physicians to conduct studies and medical investigations, and it responded very positively and helped the doctors in their venture. The suggestions given to the company by the European doctors were believed to be very important for the progress of medical science. Heyne, who learnt Tamil, was interested in the local vegetation. He engaged seriously in the study of *Kalpasthanam*, a medical text dealing with drugs.[34] This medical work by Korakkar Siddhar deals with general rules of pharmacy. Heyne mentions the difficulties he encountered in understanding the meaning of *Kalpasthanam*, which he had acquired in a Telugu translation from the original Tamil. Further, he found the local plant names very difficult to trace. According to him, the first section of the work discussed vegetable medicines. The second section related to mineral remedies and was less abstrusely written. The third section was wanting in the copy of the manuscript he owned, meaning that Benjamin Heyne did not have the opportunity to peruse it.[35]

Part one of the *Kalpasthanam* manuscript deals with climate, weather, and nature of different soils, the proper method of collecting medicines as prescribed in the *sastras*, and the manner of making nectar-like medicines. It gives a list of plants, most of which had medical uses. The parts of the plants listed included tuberous and bulbous roots, the bark of large trees, trees possessing a peculiar smell, leaves, flowers, fruits, seeds, acrid and astringent vegetables, milk plants, gums, and resins. The text also contained descriptions of types of medicine such as decoctions, pills, powders, and oils.[36]

Part two was the treatise on medicine, consisting of nine chapters. Chapter one gives advice to physicians. Chapter two deals with the pulse. Chapter three contains the diagnoses of the three principal disorders. Chapter four discusses the principal diseases that arose from *vaatham*, *pitham*, and *kapam*. Chapter five deals with the causes of diseases. Chapter seven deals with diet. Chapter eight deals with fevers. Chapter nine gives an account of fever and epilepsy. Chapter ten deals with prognosis and mentions metallic medicines, including the *paspam* of tutanag (zinc), *sinduram* of irumbu (iron), and *paspam* of thamira (copper).[37]

The *Kalpasthanam* manuscript deals, in particular, with the medicinal value of *Korakkar muli* (*Cannabis sativa*). The text also describes the medicines for fevers, and a few other disorders, besides the preparation of medicines including *vatacilerpane kunam* (medicines for diseases associated with phlegm); *puta centura navakutori marunthu* (the preparation of medicines by mixing the nine ores through a kind of heating process with a metallic oxide); *vaciyam* (subjugating potions); *perumpattukku marunthu* (medicine for menstrual problems); *navamulattirkup parpamum neyyum* (powder and ghee for nine types of piles); the reason for *kirani* (chronic dysentery), its nature, and *suranam* (powder); *kunma kutori senduram* (remedies prepared through processing nine ores that help avert indigestion); the ways to prepare medicine from salt and the medicinal value of salt; *cippiyantan sangu parpam* (the powdered medicine made out of chank lime, derived from the chank shell); the emission of acid water from the body; *irata parpatukku amuthamaka suranam* (the method of preparing medicinal powder from mercury); *tamira parpam* (calcinated white powder of copper); *makaraca pupati senduram* (decoction to cure asthma); and *makatica uppa cayanir* (medicine for the segregation of acids in the cardinal points).

Heyne shared the prejudices of many Europeans of his day, believing that the natives were generally less culturally advanced than the British and thus considering most Tamil medical practitioners to be quacks. However, he conceded that the medical knowledge of the Tamils was not altogether false. He opined that the medical works of Tamil Hindus were to be regarded neither as miraculous productions of wisdom, nor as repositories of nonsense. He believed that the practical principles on which their medical practice was founded were similar to those of the Europeans. He stated that the theories of the Hindus could be reconciled easily with those of the Europeans, adding that allowances should be made for the ignorance of the Hindus on the subject of anatomy.[38] Thus, Siddha medical knowledge written in English came to be circulated in London.

Whitelaw Ainslie was an assistant surgeon in the English settlement of Madras in 1788. When appointed as a full surgeon of the English Company in 1794, he expressed his keen desire to study the native plant drugs.[39] He mentioned that knowledge of the local vegetation was vital for medical reasons. He wrote on useful plants in Tamil country that had not yet been within the reach of scientific investigation of the Europeans. He devoted 20 years to identifying bazaar drugs that were later included in the British materia medica. His pioneering studies made an important contribution to Western understandings of Tamil medical knowledge.

Ainslie studied the Tamil medical text of *Agasthiyar Vaythiam Ayinooru* that dealt with medicine.[40] He compiled the "Materia Medica of Hindoostan" in 1813. It was sponsored for printing by the Government of Madras[41] and dedicated to his patron, the famous botanist Johann Peter Rottler.[42] Ainslie studied the plant names in several Indian languages as well as the Arabic, Persian, and Latin names he included in the work. Despite its comprehensive nature, the work received criticism from some quarters: Robert Wight, head of the botanical garden of Madras, castigated Ainslie's work as little better than a monument of absorptive labor since the book did not have illustrations. Nonetheless, Ainslie's book, initially printed in Madras, was revised and a new edition appeared in two volumes from London in 1826.[43] Thus, some elements of Siddha medical knowledge came to be spread through printing in London.

The English Debate and Controversy over the Use of Snake Pills Prepared by the Siddha Doctors

Pratik Chakrabarti has discussed the Thanjavur (Tanjore) pills used for snakebites at length.[44] Here, I would like to mention some aspects of the transmission of knowledge about these pills that have not yet come to light, and to emphasize the significance of the English debate and controversy over their use. Christian Frederick Schwartz was a Protestant missionary who worked in the Thanjavur area in 1778.[45] He was struck with wonder when he saw a Tamil doctor preparing pills for snakebites. According to him, the native doctor was also very skilled in the treatment of poisonous insect bites. Schwartz therefore called upon Dr. Thomas Strange, the English surgeon stationed in Thanjavur, and introduced him to the Tamil doctor in July 1788. Schwartz and Strange had long and fruitful discussions with the Tamil doctor and agreed afterward to write to Archibald Campbell, the governor in Madras, about the native doctor and his snake pills. Dr. Strange wrote a letter in which he recommended that the snake pills of Thanjavur be examined by the English Company doctors in Madras. He added that he desired to make the best use of the situation and affirmed that the Tamil doctor had cured several persons. The Governor replied and asked for the snake pills to be sent from Thanjavur to Madras, where the Hospital Board could conduct trials to assess the true merits of the pills.[46] The snake pills were obtained and dispatched to Madras, where a thorough analysis was carried out. William Duffin and other surgeons thereafter wrote to the Hospital Board in September 1788, stating that the results of the repeated trials were highly successful. Hence they recommended the pills to the Madras Government for the treatment of snakebites, specifying that it should be left to every doctor to administer the remedies according to his own judgment in the cases of venomous bites.[47] It was pointed out that some of the components of the snake pills had raised certain queries but clarification was left until later. The government decided to publish the details of the report. William Duffin and the other surgeons praised the government's decision to do so, since doubts regarding the ingredients had been raised by some medical experts.[48] In October 1788, after the preliminary analysis was over, the Military Board of Madras approved and encouraged the use of the pills for the treatment of snake and mad dog bites.[49] It requested the medical storekeepers to keep sufficient quantities of the snake pills of Thanjavur for supply when needed.[50] Thus we find that a simple adoption of local medicine, despite the public's approval of the pills received as a cure, was not easily accepted by the English Company administration in Madras.

Some English doctors, who prescribed the snake pills as one method of treatment, conducted further analysis of the ingredients. This became perceived as vital, since serious doubts were cast on the contents of the pills prepared by the native doctors. Without any delay, the Physician-General of Madras and the hospital members began to make an analysis of the pills. They wanted to exactly know the names of the ingredients and to draw up an accurate account of the proportions of each substance in the pills. They attempted to specify, in a particular manner, the Tamil and English names of each ingredient, as well as the botanical names of such plants or roots used in the composition of these pills, as well as the manner of using the snake pills.[51] The Military Board also wished to publish the results without delay in the *Madras Courier*, with details regarding the proper administration of the pills.[52]

James Anderson, the Physician-General at Fort St. George in Madras, produced a detailed list of the ingredients of the pills, in his minute prepared in November 1788. He added that it was indeed arsenic that put the usage in doubt. He demanded a ban on the internal use of the pills, arguing that such noxious drugs should be rejected by the British for the sake of humanity.[53]

The views of James Anderson were seriously contested by William Duffin, the Head Surgeon of the hospital in Madras. He wrote in his minute that despite their arsenic content, he had found the snake pills very beneficial in most cases. Duffin added that he himself had earlier transmitted the results to Patrick Russell, the physician and naturalist to the English EIC, who had communicated the details to the Royal Society of London.[54] Duffin defended his opinion strongly, pointing out that the English were still ignorant of the analysis of the exotic plants, and other vegetable matter that was compounded with the minerals. He suggested that ways might be found with which to counteract the toxicity of arsenic and argued that the English establishment should not hastily condemn a remedy that experience and demonstration had, in many cases, proven successful. The remedy could be given with safety; besides, the internal use of arsenic was, by no means, prohibited. He also mentioned that established practitioners in Europe prescribed arsenic in cases of cancer and intermittent fevers. Therefore, Duffin was of the opinion that the formula should be published in the *Madras Courier*.[55]

In 1789, Sir William Jones wrote to James Anderson about his own success with arsenic, mentioning that he was very interested in the details of the snake pills. He admitted frankly that until the time he saw Duncan's Medical Commentaries, he had the same abhorrence as James Anderson toward the use of arsenic in medicine. He had originally suppressed a paper by a Tamil physician about a cure of elephantiasis by small quantities of arsenic. However, Jones stated that when he found that arsenic had been safely given in England for treatment, he ventured to print the paper in the second volume of the *Journal of the Royal Asiatic Society* published from Calcutta. He stated that poisons must, no doubt, be administered with extreme caution. But if banished from practice, he asked, what would become of antimony, mercury, and opium, and without these three remedies, what would happen to mankind?[56] Jones mentioned that people should readily admit and consider the physical environment, the stimulation of the region, and the nature of their food. He added that it was very likely that chemistry had flourished in the Tamil country.[57] In his reply to William Jones, James Anderson merely urged caution in the use of arsenic or any other poison in medicine and added that Tamils were better acquainted with the use of simple medicines than any other people in the world. Thus, duality is evident in the writings of James Anderson. He also changed his opinion with regard to the use of snake pills. As was the case with William Jones, the positive comparison of Tamil with European techniques seemed to be a decisive factor in changing his views.

Patrick Russell, the English Company surgeon, temporarily put an end to the debate over the use of snake pills with his book *An Account of Indian Serpents Collected on the Coast of Coromandel*, published from London in 1796. He described his experiments with the Thanjavur snake pills as inconclusive.[58] W. Boag mentioned in 1809 that the Tamils had long employed arsenic, the principal ingredient in Thanjavur snake pills. The little experience collected by the Europeans did not enable them to form an exact judgment regarding the snake pill. One problem was that it was often difficult to distinguish the possible side effects of the remedy from the symptoms of the disease. Boag opined that it should be probably employed in desperate cases only and where no other powerful remedy could be procured.[59] Despite the concerns of some European medical men, Thanjavur snake pills continued to be used as an antidote for snake bites even in the nineteenth century. T. Lauder Brunton and J. Fayrer wrote that although the snake pills enjoyed a large amount of popular confidence, when they were tested by careful experiments in 1873 they failed to provide the desired results.[60] As the example of the Tanjore pills demonstrates, Europeans keenly observed Siddha medicine preparations, and purchased pills, tablets, and powders and they conducted in-depth investigations into the composition of various substances before using them. However, even when experience showed that the medicines were safe, they often hesitated to use them themselves.

An Assessment of the Study of Tamil Medical Manuscripts by Early Modern Europeans

During the medieval and the early modern periods, medical manuscripts in Tamil were composed only in poetry. Many of them were in short verses and some were long. Siddha medical texts were composed in the *andaadhi yaapu* (anaphora) style. The exact year, month, and date when such works were composed or completed by the poets remain unknown. However, it is not very difficult to find out the probable period based on the style of language used by the texts. Copyists were always employed to make new palm-leaf manuscripts based on the old manuscripts available. The new manuscripts were meant for circulation since printing was unknown to the Tamils before the colonial period. Later copyists often very neatly furnished their name, year, month, and day of completion of the work, sometimes with other related details, such as the place and the person who possessed the manuscript. Sometimes this was done as a result of a request from the European who commissioned the copy.

Medical manuscripts in Tamil often contained Sanskrit words that were in current usage. Several colloquial terms, such as *arachu* (grind), *kulundu* (cool), *vedichu* (burst), and *vevikka* (boil), were commonly used. Literary forms of expression, such as *thaamarai vazhaiyam* (*thandu* or stem) and *thengin ennai* (coconut oil), have also been been found in medical manuscripts. In some cases, the names of herbs given in the manuscripts were altered by changing the letters at the beginning, for example, *kuronidha suram* (*suronidha suram*), *kodisa paalai* (*kudasap paalai*), *singani naadi* (*sangani naad*), and *permati* (*peramutti*). Similarly, we also find changes in the alphabets in the middle of words, such as *koovazhai* (*koovizhai*) and *maathazhlam* (*maathuzhlam*). We find alphabets changed at the end, too, such as *karikonnai* (*karikonrai*), *thripala* (*tripalai*), *ponnaangkaani* (*ponnaangkanni*), and *milakuraiyan* (*milakaranai*). Sometimes one word is changed into another in the manuscripts, such as *kodi sappani* (*kudasapaalai*), *thiga merukkam* (*thigal erukkam*), and *mudakkotkai* (*mudakottraan*).[61] It is difficult to say with certainty, but these may have been deliberate strategies aimed at rendering the contents secret or to make them only accessible to those aware of the code language. Under these circumstances, Europeans found it extremely difficult to understand the contents of the Siddha medical manuscripts, since they had not acquired the necessary linguistic skill: small wonder given the poetic form of most texts. They therefore employed *dubashis* (linguists, literally meaning "two languages") with whose help and assistance they could comprehend the contents and compose prose versions of the texts in French, Danish, German, and English. They thereby rendered a very valuable service to European society.

In order to enable the European reader to form some judgment of the merits of Siddha medicine, the missionaries and company surgeons ventured upon translations. In fact, they desired a compilation of everything that the Tamils had written on Siddha medical science. The Europeans therefore first attempted a literal translation of Siddha works, but after many trials they were obliged to give up the task since it was beyond their capacity and power. The next alternative plan they made was to make extracts, but the style of the Tamil authors also rendered this a very difficult task. What they then wished to present to the reader was a kind of medium between these two approaches. In some cases they adhered to the Tamil authors of the Siddha manuscript very closely, while in others they abridged and omitted many quotations that appeared to them quite useless to the European reader. As a result of the poetic style abounding in similes, metaphors, and all kinds of figures of speech employed by Siddha medical manuscripts, the missionaries and company surgeons considered it impossible to make exact and faithful translations into European languages such as French, Danish, German, and English. The Tamil medical texts were, of course, replete with allusions unintelligible to the Europeans.

Conclusion

The Tamil medical texts became examples of medical knowledge of extra-European provenance to many European doctors because they noticed innumerable Siddha medical texts and became interested in reading them and collecting the relevant information. They eventually converted the same data into their own forms of knowledge in a format designed for European audiences. In order to validate the knowledge gained in this way, they also desired to verify it through experimentation. Hence, we may agree with the view of Harold J. Cook that, during the scientific revolution, physicians turned their backs on received knowledge in the quest for truth, but that over a period of time, there was a shift toward incorporating knowledge from elsewhere that had practical applicability and medical knowledge that could be traded for material profit.

Several individual Europeans studied Tamil medical texts thoroughly, often focusing closely on a particular Siddha manuscript that interested them. They made some experiments on certain herbs but could not obtain the desired results. They made comparisons and confused certain diseases that broke out in Europe with those they encountered in Tamil country. Several understood that only through the study of Siddha medical treatises could they know the secrets of successful medical practice in the region. Upon this realization, they immersed themselves in the study of Tamil language and thereafter the related medical texts. At times, they even purchased manuscripts and sent them for preservation to Europe, with a view of storing this knowledge for posterity. Thus, some aspects of Siddha medical knowledge and ideas were transferred to Europe. The Europeans were interested primarily in the medical efficacy of applied substances, an efficacy that could be exploited for healing or as a marketable commodity. Portable, profitable, and, above all, practical remedies that could be marketed and sold for application to specific maladies were in demand. Siddha medical knowledge, to a large measure, was thus imported, invented, endorsed, copied, adapted, reformulated, and marketed in Europe between 1700 and 1850. The Euro-Tamil medical dialogue had meaningfully and significantly contributed to the advancement of both cultures in the early modern world. It may be said that the Europeans had carried out their work following the orientalist tradition of reading original

Tamil texts. All of the individuals mentioned here had different backgrounds, such as missionaries, pharmacists, military surgeons, company hospital physicians, and botanists, and they developed their own particular interests while investigating Siddha medicine. However, a general attraction to toxicology and an interest in the use of toxic substances such mercury, sulphur, and arsenic in the Tamil coast, are notable.

The Europeans who attempted to collect information on Tamil medicine and details from the native physicians could not collect all that they required through conversation and dialogue. Tamil doctors were often unwilling to pass on all the professional secrets of medicine. European doctors were highly inquisitive, curious, and all the more skeptical of what they heard orally. The number of questions posed and the answers given by Tamil doctors were not at all satisfactory to the wishes of the Europeans. The information they obtained had to be verified and ascertained through the study of Tamil medical treatises. Some Europeans who studied Siddha medical texts perceived Tamil country to be the cradle of medical knowledge. However, their prejudices led them to believe that the knowledge possessed by the natives then living there had been clouded by Hindu religious "superstitions." Europeans tended to assume that they alone could extract the valuable scientific and medical truths from Siddha texts because they utilized recently discovered European methods. They were mainly interested in studying plants' constituents, pharmacological aspects, active principles, and materia medica. They learnt about the dosages of medicine, to be administered internally and externally, and rules of diet, including the prohibition of certain foods. Thanks to the works of innovators such as Andreas Vesalias and William Harvey, during the early period of the EIC's commercial contact and expansion, Europeans gradually moved away from a humoral view toward a chemical or mechanical view of the human body, which changed the medical notions of the Europeans on a basic level. Nonetheless, many important findings of Tamil medicine were subjected to European examination and analysis before the data was transferred to Europe for either acceptance or rejection.

Company physicians and surgeons were, in the early days of the European presence in the Tamil coast, at a loss to understand and treat the various diseases that were peculiar to the climatic and weather conditions. Many physicians of the English Company did not venture into the study of the Ayurvedic, Siddha, and Unani systems, but instead developed their own treatments and explanations for disease. James Lind, a physician in the naval hospital of Fort St. George in Madras, wrote and published on tropical diseases in 1768. He characterized tropical diseases as "diseases of strangers" and thought they were similar all over the world. He mentioned examples of higher mortality in hot climates, and drew the attention of all the commercial nations of Europe toward the important objective of preserving the health of their countrymen, whose business carried them beyond seas.[62] Since then, several writings appeared linking tropical medicine, imperialism, and the culture of scientific racism.

Geographically speaking, Asia or India is a land of tremendous diversity of human cultures, with different agro-climatic and vegetative zones. These cultures exhibit diverse technologies of resource use and of social modes of resource control. Hence, it may not be proper to talk of tropical medicine in general, and we need to look for particularities such as Ayurveda, Siddha, and Unani to understand in depth the medical systems within regions such as Tamil country. Europeans looked for Siddha medical texts for a broad reading on health, disease, and medicine. Siddha texts contained no idea of the environmental theories of disease and did not demonstrate a racial difference. Europeans also did not see the perceptiveness of tropical medicine clearly, and as noted by Michael Worboys, the knowledge of parasitology, bacteriology, and virology had not yet developed in Europe.[63] The growing sense of distinctiveness reflected in British medical systems therefore proved a barrier to the transmission of Siddha medical ideas and practices from the Tamil coast to Europe in the late nineteenth century.

Notes

1. Harold J. Cook (2007), *Matters of Exchange: Commerce, Medicine and Science in the Dutch Golden Age* (New Haven, CT: Yale University Press).
2. David Arnold (1993), *Colonizing the Body: State, Medicine and Epidemic Disease in Nineteenth-Century India* (Berkeley: University of California Press); David Arnold, ed. (1996), *Warm Climates and Western Medicine: The Emergence of Tropical Medicine, 1500–1900* (Amsterdam: Rodopi), pp. 7–8.

3. Pratik Chakrabarti (2011), *Materials and Medicine: Trade, Conquest and Therauptics in Eighteenth Century* (Manchester: Manchester University Press).
4. The date and provenance of the medical text of *Sillarai Kovai* is unknown. A palm-leaf manuscript is available at the library of the Institute of Asian Studies, Chennai, India. A. Thasarathan and V. S. Subbaraman, eds. (1993), *A Descriptive Catalogue of Palm-Leaf Manuscripts in Tamil*, 3 vols. (Madras: Institute of Asian Studies).
5. Gaston Laurent Coeurdoux was born in Bourges on October 18, 1691 and joined the Society of Jesus as a novice in 1715. He left France for the Madurai Jesuit Mission in 1732. He first served at the French settlement in Pondicherry and in 1740 became the Jesuit superior at Karaikal. In 1745, he was appointed as the procurator in Pondicherry. In 1747, he became the Jesuit superior there until he was replaced by Fr. Lavud in 1751. Coeurdoux continued his religious work in Pondicherry and died in 1779 at the age of 83. L. Besse (1918), *Liste alphabétiques des missionnairs du Carnatic de la Compagnie de Jésus au XIIIe Siècle* (Trichinopoly: St. Joseph's Press), p. 27. L. Besse (1780–1794), *Lettres édifiantes et curieuses, écrites des missions étrangères par quelques missionaires de la Compagnie de Jésus*, Paris: Vol. 28, Nouvelle Editions, Vol. 14, pp. 97–121.
6. Madurai Province Jesuit Archives, Shenbaganur, India, *Documents de la Mission du Carnatic*, Vol. 3, Document No. 48, ff.159–160, letter of Fr. Coeurdoux dated September 29, 1738.
7. Friar Jean-Baptiste du Choiseul was born on January 31, 1717, entered the Society of Jesus on March 28, 1737, and arrived in Pondicherry in 1740. He was appointed in charge of the Jesuit pharmacie in 1742. Choiseul died in Pondicherry on July 19, 1793. L. Besse, *Liste Alphabetiques des Missionnaires du Carnatic*, p. 27.
8. S. Jeyaseela Stephen (1998), *Portuguese in the Tamil Coast: Historical Exploration of Commerce and Culture, 1507–1749* (Pondicherry: Navajyothi), p. 312; Adrian Launay (1898), *Histoire des Missions de l'Inde—Pondichéry, Maissour, Coimbatour*, Vol. 1 (Paris), p. 219.
9. Madurai Province Jesuit Archives, Shenbaganur, India; Microfilm, no. 8.
10. Claude du Choiseul (1756), *Nouvelle méthode sure, courte et facile pour le traitement de personnes attaqués par de la rage* (Paris: H. L. Guerin & L. F. Delatour).
11. Bartholomäus Zeigenbalg was born in Pulsnitz in Lusatia (Saxony) on June 24, 1683; he studied at Halle and was ordained as a missionary. He left Copenhagen in a ship that sailed on November 29, 1705 and arrived at Tranquebar (after 222 days of voyage) on July 9, 1706. He worked in the Danish mission and died at Tranquebar on February 23, 1719 and was buried on February 24, 1719 in the New Jerusalem Church. Johan Ferdinand Fenger (1863), *History of the Tranquebar Mission: Worked Out from Original Papers*, trans. Emil Francke (Tranquebar: Evangelical Lutheran Mission Press), p. 312.

12. C. S. Mohanavelu (1993), *German Tamilology: German Contributions to Tamil Language, Literature and Culture during the Period 1706–1945* (Madras: Saiva Siddhanta Pathippu Kazhagam), p. 123.
13. Archiv der Franckeschen Stiftungen (AFSt), Halle, Tamil Manuscripts, No. 112, *Waguda Tschuwadi*.
14. W. Caland (1926), *Ziegenbalg Malabarische Heidenthum: Herausgegeben und mit Indices* (Amsterdam: Koninklijke Akademie van Wetenschappen).
15. William Germann (1880), "Ziegenbalg's Bibliotheca Malabarica," *Missionsnachrichten der ostindischen missionsanstalt zu Halle*, 32 (3–4): 91.
16. Ibid. p. 84.
17. AFSt, Halle, Tamil Manuscripts, No. 57, *Udelkuddu Wannam*; Germann, Ziegenbalg's Bibliotheca Malabarica, p. 84; Arno Lehmann (1955), "Hallesche mediziner und medizinen am anfang deutsch-indischer beziehungen," *Wissenschaftliche Zeitschrift, Martin-Luther-Universitat, Mathematische-Naturwissenschaftliche Reihe*, 5, 117–132.
18. Johann Ernst Grundler was born in Weissensee on April 7, 1677; he studied at Halle, was ordained at Copenhagen in 1708, and sailed in a ship that left Copenhagen on January 17, 1708. He landed at Tranquebar and moved to Poraiyar and lived there and died on March 19, 1720. He was buried the next day in the New Jerusalem Church at Tranquebar. Fenger, *History of the Tranquebar Mission*, p. 312.
19. *Der Königl. Dänischen Missionarien aus Ost-Indien eingesandter ausführlichen Berichten, Von dem Werck ihrs Ams unter den Heyden, angerichteten Schulen und Gemeinen, ereigneten Hindernissen und schweren Umstanden; Beschaffenheit des Malabarischen Heydenthums, gepflogenen brieflicher Correspondentz und mundlchen Unterredungen mit selbigen heyden* (1710–1772), *Hallesche Berichte* Waiserihaus, Halle, Teil 1–9, Continuationen 1–108, Teil 1, pp. 286–287.
20. AFSt, Halle, Tamil Manuscripts M 2, B 11.
21. *Hallesche Berichte*, Teil 1, pp. 286–287.
22. AFSt, Halle, Tamil Manuscripts M 2, B 11, fls. 6–19.
23. Ibid. fls. 29–35.
24. Ibid. Catalogue S. 120–132, S. 133–149.
25. Sarasvathi Mahal Library, Thanjavur, India, *Tamil Medical Manuscripts*, No. 14; S. Venkata Rajan, ed. (1980), *Agathiyar Irrandayiram* (Thanjavur: Don Bosco Press).
26. Mohavavelu, `, pp. 43–45.
27. Theodor Ludvig Frederich Folly was born in 1740 in Itzehoe, a town in the Duchy of Holstein in the northern part of present-day Germany. In 1769, he was employed as second surgeon on one of the ships of the Danish Asiatic Company. He also worked in the dispensary of the Royal Frederik's Hospital in Copenhagen. He was employed later as second surgeon in the ships of the Danish Asiatic Company from 1770 to 1775 and sailed thrice in the voyages from Copenhagen to Tranquebar. In 1777, he was appointed as second surgeon in the military hospital of Tranquebar; he left Copenhagen in 1778 and arrived at Tranquebar in 1779. He worked along

with Gottlieb Friderich Böttger, the head surgeon, from 1779 to 1786. Folly became the head surgeon in charge of the military hospital in 1786 and later died in Tranquebar on July 30, 1803. See Niklas Thode Jensen (2005), "The Medical Skills of the Malabar Doctors in Tranquebar, India, as Recorded by Surgeon TLF Folly, 1798," *Medical History*, 49 (4): 489–515.
28. Det Kongelige Bibliotek, Kobenhavn, Mss. No. Add. 761a, tillage 4o; T. L. Folly, *Bemerkungen der von Dr. Roxburgh entdeckten Fieber-Rindse Swietenia febrifuga*, 1792, fls. 1–18; See also Mss. No. Add. 761e 8o; T. L. Folly, *Nogle Anmaerkninger I Anledning af Doctor Kleins forsog omden veneriske syges helbredelse I ostindien, Tranquebar, 1798*.
29. Det Kongelige Bibliotek, Kobenhavn, Rare Manuscripts Section, Mss. No. Add. 761e 4o.
30. Kirurgisk Akademi, Kobenhavn, Mss. No. Add. 333a.
31. Jensen, "The Medical Skills of the Malabar Doctors," 503–504.
32. Benjamin Heyne was born in Dobra in Germany in 1769. He reached Tranquebar in 1792 and was later appointed as a botanist in Samalkota in 1794, and then became an assistant surgeon in Madras in 1799. He went on leave to Europe from 1812 to 1815. Heyne returned to Madras and continued his medical practice there. He died at Madras in 1819. See Natural History Museum (hereafter NHM), Department of Botany Library, London, Mss. Rox; the letter of Christopher John written from Tranquebar, dated July 25, 1793, to William Roxburgh; and the letter of Christopher John and the Moravian Brothers written from Tranquebar, in August 1793, to Andrew Ross.
33. Tamil Nadu State Archives (TNSA), Chennai, India, *Surgeon General's Records* (SGR), Vol. 12, fl.53.
34. Benjamine Heyne (1814), *Tracts: Historical and Statistical on India with Journals of Several Tour through Various Parts of the Peninsula also an Account of Sumatra in a Series of Letters* (London: Black Parry), p. 126.
35. Ibid., p. 126.
36. Ibid., pp. 127–144.
37. Ibid., pp. 149–167.
38. Ibid., p. 165.
39. Whitelaw Ainslie was born in 1767. He joined the English Company's service and became an assistant surgeon on June 17, 1788 and served at the Chingleput garrison. He was promoted to a surgeon on October 17, 1794. He was appointed as the superintendent surgeon of the southern districts of the army of Madras in 1810 and remained in the post till 1814. In the next year, he resigned from the job and returned to England. He died in 1835.
40. Bibliothèque Nationale, Paris, Mss. Indien (Tamoule), No. 111, *Agasthiyar Vaythiam Ayinooru*, fls. 1–82.
41. Whitelaw Ainslie (1813), *Materia Medica of Hindoostan…* (Madras: Government Press).
42. The text runs as follows: "To Reverend Doctor J.P. Rottler, the following pages are inscribed, as a token of respect for his scientific celebrity, of

esteem or his private character and in grateful acknowledgement of the kind and liberal aid which has been received from him, by his sincere friend Whitelaw Ainslie."

43. Whitelaw Ainslie (1826), *Materia Indica: Or Some Account of Those Articles which Are Employed by the Hindoos and Other Eastern Nations in their Medicines, Arts and Agriculture*, 2 vols. (London).
44. Pratik Chakrabarti (2006), "Neither of Meate nor Drinke, but What the Doctor Alloweth: Medicine amidst War and Commerce in Eighteenth-Century Madras," *Bulletin of the History of Medicine*, 80 (1): 1–38.
45. Christian Frederick Schwartz was born on October 26, 1726 in Sonnenberg in Brandenburg, studied at Halle, was ordained in Copenhagen in 1749, and left for India and reached Cuddalore on July 30, 1750. He worked as a missionary for 11 years at Tranquebar, established the Tiruchirapalli mission in 1762, and later settled down at Thanjavur in 1778. He continued his work and died there on February 13, 1798. Fenger, *History of the Tranquebar Mission*, pp. 312–313.
46. TNSA, SGR, Vol. 3, fl.175; see the letter from the Military Board to James Anderson, the Physician-General and Members of the Hospital Board, dated September 17, 1788.
47. Ibid., fls.185–186; see the letter of the Hospital Board dated September 26, 1788.
48. Ibid.
49. Ibid., fls. 223–225,;see the letter from the Military Board to James Anderson, the Physician-General and Members of the Hospital Board, dated October 22, 1788.
50. Ibid.
51. Ibid.
52. Ibid., fl. 224.
53. Ibid., fls. 230–231; see James Anderson's opinion on the pills, dated November 10, 1788. See also Chakrabarti, "Neither of Meate nor Drinke," p. 20.
54. Ibid., fls. 238–241; see Duffin's opinion communicated to the Hospital Board, dated November 17, 1788.
55. Ibid., fl. 241.
56. NHM, Department of Botany Library, London, Mss. Rox; see the letter of William Jones from Crishnanagar, dated September 14, 1789, to James Anderson at Fort St. George.
57. Ibid.
58. Patrick Russell (1796), *An Account of Indian Serpents Collected on the Coast of Coromandel; Containing Descriptions and Drawings of Each Species; Together with Experiments and Remarks on Their Several Poisons* (London: W. Bulmer Shakespeare Press), pp. 76–77.
59. William Boag (1799), "On the Poison of Serpents," *Asiatic Researches Comprising History and Antiquities, the Arts, Sciences and Literature of Asia*, 6: 112–113.

60. T. Lauder Brunton and J. Fayrer (1873–74), "On the Nature and Physiological Action of the Poison of Naja Tripudians and Other Indian Venomous Snakes—Part II," *Proceedings of the Royal Society of London*, 22: 132–133; Oliver Phelps Brown (1890), *The Complete Herbalist; or, the People Their Own Physicians...* (London: W. Bulmer Shakespeare Press), p. 455.
61. Sarasvathi Mahal Library, Thanjavur, India, *Tamil Medical Manuscripts*, No. 14, fls. 9–14, 23, 27–38, 43–51; Bibliothèque Nationale, Paris, *Mss. Indien (Tamoule)*, No. 111, *Agasthiyar Vaythiam Ayinooru*, fls. 7, 29, 42, 57–59, 66–80.
62. James Lind (1768, 1792), *An Essay on the Disease Incidental to Europeans in Hot Climates*, 1st ed., 5th ed. (London: J. Murray), pp. 7–8, 154.
63. Michael Worboys (1996), "Germs, Malaria and the Invention of Mansonian Tropical Medicine: From 'Diseases in the Tropics' to 'Tropical Disease,'" in Arnold, *Warm Climates and Western Medicine*, pp. 188–193.

5

Medicine, Money, and the Making of the East India Company State: William Roxburgh in Madras, c. 1790

Minakshi Menon

In a fascinating revisionist study of the English East India Company (EIC) in seventeenth-century India, the historian Philip J. Stern has pointed out how the EIC was by its very organization a government in its own right, deserving analysis on its own terms. He calls it the *Company-State*, a formation that came into being as part of an early modern empire that was itself constituted through sets of overlapping and competing political forms, of which the EIC was one.[1] Stern reads against the grain of existing historiography on the EIC, which, he says, has led scholars to imagine its politics "as a subset of seventeenth-century English and European politics, political economy, and state formations."[2] Instead he argues that the EIC's constitution was volatile, and allowed it a protean existence by balancing various forms of authority, leading it sometimes to claim that it was a "mere merchant" and at others an "independent sovereign."

Stern reveals continuities between the seventeenth-century EIC and its eighteenth-century incarnation as a territorial power. Both forms governed—administered law, collected taxes, provided protection, inflicted punishment, "enacted stateliness." Yet he does not tell us how the EIC's commercial practices affected its governance in the seventeenth century. What made the hyphenated entity, the *Company*-State, a distinctive political economic order?

What aspects of its commercial being organized its stateliness? This chapter takes up these questions in the context of Company state making in Madras in the late eighteenth century.

It is usual to speak of the EIC state as a single entity, but the process of its construction varied in its three principal settlements: Bengal, Madras, and Bombay. At Madras, the relationship between Company and other elites was determined by the tortuous nature of the public debt, which bound the settlement's governor and council ever more closely to its free merchants, and their own servants acting in a private capacity.[3] This was expressed in the synergy between Company business and private trade, which helped the upper levels of the state cohere. Asserting political sovereignty required commercial cooperation between Company personnel and independent actors, which was reflected in the creation of a new state institution.[4]

The chapter makes its arguments by examining the Indian career of the eighteenth-century Scottish surgeon and botanist William Roxburgh (1751–1815) and his relationship with his chief patron in India, the Madras free merchant Andrew Ross (d. 1797). Roxburgh's medical training at the University of Edinburgh equipped him with the skills required to make natural knowledge useful for commerce. He was able to exchange those skills for Ross's patronage, which, in turn, created opportunities for upward mobility within the Company state, and built him a fortune through private trade in the Indian Ocean World (IOW). It also gave him the power to shape one of the emerging scientific institutions of the colonial state, the office of Company Naturalist. I explore two themes that are analytically separate, but worked together in this example of career building in the Company state: the role of "familial" ideology and the place of "logistical" knowledge in EIC state making.

Roxburgh's relationship with Ross, as revealed through the letters Ross wrote to him, had a familial character. The letters could as well have been written by a caring paterfamilias to a loyal son, as a patron to a client, illustrating an instance of the expansion of notions of paternity into the polity that the sociologist Julia Adams has argued underlay early modern European state building: "Paternity was interpreted through the optic of official ideologies of masculine activity, creativity and power that extended beyond lineal reproduction to a more general sense of political husbandry and direction."⁵ In accepting Roxburgh as a client, Ross was developing a relationship that would increase his grip on Indian Ocean commerce, but he was also helping Roxburgh build a career and acquire a patrimonial stake in the Company state, by directing his political fortunes. In return, Roxburgh placed his natural-knowledge-making abilities at his patron's disposal.

The natural-knowledge-making techniques taught to aspiring medics⁶ at Edinburgh emphasized the importance of an accurate descriptive knowledge of nature and of giving full play to the powers of induction in the service of the useful arts. These aims were manifested, for example, in Dr. John Walker's natural history lectures where any account of a mineralogical or botanical object had to include its agricultural, pharmaceutical, or industrial uses, its "useful" natural history.⁷ When discussing mineralogy, Walker was at pains to present his students with examples of particular soils, places in Britain where they could be found, and the uses to which they were put:

> *Ordo I. Figulino* ~ *Figuline Earths* Is the common name applied because they have been used by Potters for their Pottery in all ages.~ *Genus Ist Argilla Clay* Is the first genus and comprehends all the common sorts of Clay. The principal and indeed its leading characters (in which it differs from others) is this, that it hardens in the fire ~ It is likewise viscid and ductile in water ~ Pott found it an Earth when pure perfectly apyrous, but by mixing other Earths with it, is easily altered ~ Here is a specimen, which is the *Argilla* (*Leneargilla* of Linnaeus); it is perfectly white and is used in Liverpool for making China[.]⁸

The same was true of Dr. William Cullen's lectures on chemistry, in which he taught the elements of chemistry through their application to the arts, and to agriculture in particular, at once addressing the needs of students of physic and of nonstudent attendees such as local improvers.⁹ This was a characteristic of the knowledge-making associated with the social relations of science in Edinburgh, where improving landlords and literati were allied in their pursuit of capital accumulation and social power.¹⁰ Edinburgh's medical savants identified socially and intellectually with their aristocratic patrons, legitimizing their claims to landed property by teaching a vision of nature that required an educated elite to understand its principles and make the best use of its resources. In practical terms, this meant that medical students taking the classes necessary to obtain their MDs were exposed to forms of knowledge and styles of knowledge-making that exceeded the requirements of the average medical practitioner of the time.

The historian of medicine Harold Cook, describing a different convergence in the Dutch Golden Age in which early modern medics and merchants both displayed a deep interest in natural facts, reminds historians of science to take note "that values of systems of *accumulation* and, particularly, of *exchange*, also changed the kind of knowledge produced."[11] Cook's examples include the development of new experimental techniques and discoveries in seventeenth-century anatomy that arose from saleable "industrial" secrets such as a receipt for embalming bodies developed by a minor lord, Louis de Bils, whose brothers were merchants in Rouen; and Leiden professor Sylvius's development of the theory of acid or alkaline aqueous substances un-fortuitously close to the bleaching fields of Haarlem.[12] Thus it was no accident, writes Cook, that the new science was born at the same time as the first global economy: "As commercial cities and the finance capital they produced became ever more important for the larger political systems of which they were a part, the values of the urban merchants, including their intellectual values, were increasingly dominant throughout society."[13] Simultaneously, says Cook, practices for making empirical knowledge of nature traveled rapidly with expanding commerce in ways that theories about nature, which were imbued with local cultural values, could not.

There are similarities between the forms of knowledge canvassed by seventeenth-century Dutch physicians and merchants, and the members of the landlord-literati alliance in Enlightenment Edinburgh.[14] Edinburgh medics' propensity to make knowledge in improving contexts carried a commitment to natural-fact making that was easily adapted to the requirements of the Company state. This was evident once the patronage politics that came into play after the Treaty of Union of 1707 gave such physicians and surgeons a majority of the posts in the EIC's Medical Service.[15]

EIC functionaries' ability to exercise strategic power within the colonial state depended on having a strong grasp of its commercial affairs. Various types of knowledge were needed for this. Factors had not only to know about the structure of demand for commodities in European markets, but also keep track of inter-Asian commerce in the southeastern Indian Ocean, as well as the state-of-play in markets in the Indian hinterland. Intimate knowledge of the sites of production and techniques of manufacture of tradable goods—agricultural products, raw materials, or finished manufactures—was key. So was knowledge of the migration and movement of different groups of indigenous merchants, such as the *banjara* nomads, inland traders who brought cotton and yarn from the Deccan to weaving centers in the Coromandel. And most importantly, Company factors had to have the expertise to judge the quality of the goods involved in their trade: silk, cotton, indigo, spices such as pepper or cardamom, bazaar medicines, and minerals. It was here that medics (who had the right to trade privately) had an edge. Their training in natural history allowed them to identify plants and minerals in the interior of the country, which they traversed with the Company's armies, and the rivers and waterways down which goods could be transported to the coast. In other words, they were masters of what Arjun Appadurai has called "production knowledge," technical knowledge of a commodity, combined with knowledge of the market, the consumer, and its destination.[16] The extent of their involvement in trade is hinted at by Governor-General Cornwallis's Minute of 1788 on the Indian Medical Service, and a memo of June 1838 submitted to the Government of Bengal.[17]

The sort of natural knowledge that Roxburgh controlled has been theorized by the sociologist Chandra Mukerji as logistics or logistical power, "the use of [the] material world for political effect." As she argues: "The exercise of logistical power depends on natural knowledge (*techne* and/or *episteme*), either practical experience in working with materials, or formal knowledge useful for reshaping the environment."[18] Mukerji's aim is to distinguish the ability to control the natural world, as a form of power quite distinct from the strategic exercise of will for domination. It shapes human action differently and acts as a distinct and independent historical force, as she shows with her example of the building of the Canal du Midi.[19]

I propose that Mukerji's theory is a valuable way to understand the power that medical men wielded when they used their training to identify or create natural objects useful for governance and trade. Such power could help produce outcomes, for example, in state making, which those without it could not achieve. Roxburgh's logistical knowledge not only increased his social and political weight, but also showed him how to use the entanglement of Company business and private trade to his benefit.[20]

Covenanted officers in Company service received abysmally low salaries. At a time when the span of an Indian career was expressed in the adage "two monsoons are the age of a man," young hopefuls did not head east for what the Company paid them. They were lured there instead by the prospect of riches acquired through independent trading in Asia. Entrepreneurial medics, such as Roxburgh, knew just how to insert themselves into the interlaced networks of Company and private trade, so that their logistical knowledge became indispensable to higher-ups in building their private fortunes. This proved particularly effective in a context where the Company's attempts to develop its sovereignty came together with its commercial functions in "shared rule," which allowed private merchants in Madras Presidency, such as Ross, to wield power in state building. It meant, in this instance, that a Madras free merchant could successfully propel his EIC client upward through the Company's ranks and into Britain's first scientific circles.

* * *

William Roxburgh, a member of an undistinguished Ayrshire family with connections to the Boswells, attended Edinburgh University in the late 1760s and early 1770s, when its Medical School was perhaps the best in Europe, and its naturalist savants were part of a "a public institutionalized alliance" that brought together landowners and literati in the interests of Scottish improvement.[21] In pedagogical terms, the alliance resulted in agronomy being treated as the "pattern science"[22] at Edinburgh, in which the developing disciplines of natural history and chemistry were embedded. Natural history was taught as of a piece with improving agriculture by John Walker, and John Hope, Professor of Botany. As the university's medical students went out into the world, they were ideologically attuned to the need to steward the land and its resources and make the knowledge required for its betterment.

After arriving in Madras Presidency in 1776, Roxburgh was first an assistant surgeon at the General Hospital at Fort St. George and then a surgeon at a small garrison town called Samulcottah (Samalkota in modern Andhra Pradesh). He remained there through the 1780s and 1790s, accepting the post of Superintendent of the Calcutta Botanic Garden in 1793. His time in Madras overlapped with the governorship of Sir Archibald Campbell of Inverneil (1786–1789). Campbell had electoral ambitions at home that he nursed on the Madras frontier by increasing the number of Scots in the EIC's civil service, creating conditions that would prove propitious for men such as Roxburgh to advance themselves in the 1790s. The large number of Scots at Madras meant that Scottish patronage politics could flourish in the Presidency at a time when its administrative culture was very fluid.[23]

EIC government in Madras was carried out through "shared rule," an analytic developed by Thomas Ertman to explain how state policy in Britain was devised and implemented through relations of collaboration (rather than coercion) between monarch and parliament.[24] In practice, this meant that administration could remain informal and susceptible to outside advice, keeping the boundary between state and society fluid. An excellent example is the role played by Sir Joseph Banks, President of the Royal Society, who became Roxburgh's most influential protector in London. Banks never held formal public office but his success as an informal advisor to a variety of government agencies and institutions saw his elevation to the Privy Council in 1797.[25] In early modern Madras, shared rule meant that all matters were decided through debate and consensus and recorded in "Consultation Books," which were constitutive of the decision-making process. It was only once a decision was written down (and read) that it became government fiat.[26] These books also signaled to EIC Directors which of their employees enabled the smooth functioning of its commercial-ruling apparatus and deserved promotion within the company's hierarchy. An up-and-coming man would have wished to appear often in them, signaling his usefulness in EIC decision-making.

Shared rule in the Madras Presidency allowed free merchants, men like Ross who were not part of the EIC's bureaucracy, to play an important role in its decision-making. Ross was the most powerful of the Madras free merchants and his reach was high. He was a sometime mayor of Madras and foreman of the Madras grand jury; he had easy access to members of the Madras Board of Revenue and was an intimate of its president, Governor Charles Oakeley, as well as of Governor-General Cornwallis at Calcutta. Tellingly, it was on his advice that Archibald Campbell had sent EIC troops to back up the expedition to establish settlements on Pulo Penang (Prince of Wales Island), so vital to the China trade.[27] And it was to him that the supporters of Prince Serfoji of Thanjavur (later Serfoji II) turned, during a succession dispute in Maratha Tanjore. Modi documents preserved in the Thanjavur Sarasvati Mahal Library describe how the Serfoji faction prepped Ross on the genealogy of the Tanjore Marathas so that he could make a convincing case for the young prince's pretensions to Lord Cornwallis.[28]

Ross's clientelist politics involved numerous Scots in Madras, especially the medical men whose contacts in British scientific circles were useful in keeping him abreast of developments in natural history that affected his trade. As a "new man" Roxburgh needed "paternal" guidance to make his way in the world; and as Ross willingly supplied it, he was ready to reciprocate by placing his logistical knowledge at his patron's disposal.

Ross's fortunes depended on staying one step ahead of emerging demand in European and Southeast Asian markets. This meant receiving information from both EIC personnel in London and merchants plying the shipping lanes of the Indian Ocean that he could translate into orders placed with his brokers. Among the items he invested in heavily were cloth, indigo, and pepper. As a coast-bound merchant, he was dependent on his informants in the interior of the country to apprise him of seasonal production in cotton and pepper, assess their quality, and provide news of political events that could disrupt the production or distribution of commodities. Roxburgh, operating from Samalkota, was one of the most reliable, and as a naturalist experimenting with dyestuffs and commercial plants in the EIC's botanic garden in the Circars, he was also able to proffer expert knowledge on the possibilities of growing newly valuable plants in the area.

Two letters from Ross to Roxburgh, written between June and December 1788, show the entwined nature of EIC business and private trade and the texture of their patron-client bond.[29] In one, Ross is busy looking for books to replace those lost by Roxburgh in the hurricane of the previous year ("I have looked repeatedly for Botanical Books since you suffered the loss of yours"), without much success ("I found the Botanist and Gardeners New Dictionary by Wheeler...I regard the title merely—and if you find it of use I shall be much pleased."), but it is on other matters that he writes—he is eager for news of Roxburgh's success with his pepper plantation. Pepper is likely to become an article of consequence, its value enhanced by the Mysore ruler Tipu Sultan's measures to keep it out of the Company's reach. Therefore, Ross urges his client:

> So you will please give me the most particular report you can of your success & before Oct^r...I think it will not be amiss if you send me Musters to be sent home to my friend [Alexander Dalrymple] for the inspection of the Directors; nor will you be displeased to hear that I sent him by the Ravensworth in Oct^r...the little Bag of Pepper which you sent to me some time before & in this Season we shall hear what they say of it...So pray exert yourself earnestly in it.

Indigo is proposed as another article grateful to trade, this time to Ross's private trade:

> It is long since I have had it in my mind to write to you about Indigo, but neglected it. I now send you 2 papers on the subject & a muster of a good quality made at Bengal. You will make every & the earlyest enquiry about it in your parts—& likeways at a distance—especially to the Westward as far as you have opportunities & acquaint me with the result. Whether it is or may soon be cultivated & made of a tolerable good quality—where—in what quantity, & at what expence, & if you meet with any send me musters—the sooner the better. If a proper attention is given to it may be rendered of consequence & therefore I wish it to rest with you.[30]

The letter of December 23 moves on to discuss cotton, and this time Roxburgh's interests are specifically mentioned: "I wish you to inquire whether Cotton of a good quality & in any quantity worth attention can be procured with (sic) Country about or the Northward of you at such a price as would answer for China."[31] He had received news from Bengal of the high prices currently paid for cotton from India in China. Could cotton, perhaps, be grown in the Circars? Had Roxburgh experimented with cotton seeds? Was labor cheap enough there to make it profitable? "You know what success the people at Bengal have had in this way lately—and to the more advantage than from Bombay—& why you (who are so indefatigable and prudent) should not try your industry in this way is not known to me."[32]

Ross indicates the way forward, commanding his man to do his bidding and Roxburgh must follow. The two were locked together in a trading partnership,[33] so there was monetary benefit to Roxburgh in doing as he was bid. But Ross's patronage could do more: he was willing to act as a broker for Roxburgh by presenting his musters of pepper to the Court of Directors through a significant intermediary, the Company's hydrographer, publisher, and propagandist, Alexander Dalrymple. Dalrymple was an important player in that part of the British public sphere where EIC doings were carefully watched and discussed. Between 1791 and 1793, he published the four numbers of the *Oriental Repertory* Volume I. The *Repertory* was meant to keep the Company's profile high at home at a time when EIC stocks fluctuated alarmingly with every new report of war in India. About half of the first number, at Ross's suggestion, was devoted to puffing Roxburgh's discovery and cultivation of pepper and indigo in the Circars.[34]

A better example of the construction of natural historical knowledge in keeping with the "values of systems of accumulation and exchange" than Roxburgh's new species of *Nerium*, the source of a very beautiful indigo dye, would be hard to find. The sharp-eyed doctor, while working on his herbarium, had noticed that its leaves contained coloring matter:

> The colour the leaves sometimes acquired, in drying for my *Hortus siccus*, first induced me to think, They were possessed of colouring matter, and the result of some Experiments, fully answered my Expectations, although I have been often deceived by the same appearances in the Leaves of other plants.[35]

It could be said, the Doctor continues, that the Company was already in possession of a sufficient number of good blues. His new indigo was, however, a marvel of easy cultivation, unlike its well-known and common cousin, *Indigofera tinctorium*,[36] and had, besides, an interesting polyvalence: it could assist the Company in stewarding its possessions while producing a profitable commodity:

> I have been as minute on every point as my knowledge of the subject permits, with a view to encourage others to undertake the manufacturing of this kind of Indigo in preference to the common kind. It is infinitely more profitable, and (what may seem paradoxical to assert) employs infinitely a greater number of hands in gathering, and bringing the leaves, cutting fuel, & c. which with every man of feeling will be an object in a country where thousands of poor miserable creatures are constantly in a state of starvation for want of employment.[37]

Medicine, Money, and East India Company State   161

The Company, meanwhile, was desirous of stabilizing its position in the Madras territories at a time of intense social and political turbulence. Developing institutions of governance and extending them into the hinterland was a major challenge.[38] The process was captured in a minute written in February 1793 by Charles Nicholas White, a member of the Board of Revenue at Fort St. George, concerning the nature of the revenue settlements to be made in a portion of the Northern Circars, Guntur district. Local elites, the zamindars, village officials, and interested British officers were equally complicit in preventing the collection of information about agrarian resources and local customs needed to stabilize revenue collections. The solution appeared to be the appointment of British district officers known as collectors, as had been done in Bengal. The current system of merchant chiefs and councils allowed no separation of the duties of revenue and commerce, which was key:

> The chiefs and councils do not transmit their accounts and vouchers in a distinct and regular manner, or enter into a simple and clear detail of the collections, or give that satisfactory elucidation, which must have been intended and expected from the institution of collectors. The latter by residing in the districts, have it also in their power to watch the conduct of the zamindars, and to check any improper designs which the chiefs and councils cannot so well attempt, from being ignorant of what is actually passing in the different zamindars' (sic).[39]

While these plans were in train, other means of collecting information at the local level were sought. Roxburgh, who was settled near the western hills of the Circars, was urged by his masters to enter the political economy of the region, dominated by *pālaiyakkārar*, "little kings," who presided over vertically integrated polities controlled through strong ties of kinship. *Pālaiyakkārar* mobilized resources through plundering each other's territories, and ensured the loyalty of their retinues by redistributing the spoils through a system of gift-giving.[40] An entry in the Madras Revenue Department Proceedings for 1793 expresses the hope that Dr. Roxburgh's "philosophic temper" would mend the manners of the hill *pālaiyakkārar* and allow their integration into the emerging Company state.[41] But the story is better told in a letter written by Ross to the Board of Revenue, supporting Roxburgh's application to rent land in

Corconda (Korukonda), near the Rampah *pālaiyakkārar's* territories in the hills.[42]

The letter, written toward the end of Roxburgh's time in the Northern Circars, is a précis of Roxburgh's accomplishments at a time of political uncertainty and Company state formation. Ross's client, we learn, is busily at work embedding the EIC state in local society while also furthering its commercial interests. He has pursued the cultivation of indigo, sugar, "new Fever Bark (*Swietenia febrifuga*), breadfruit, cinnamon and coffee, and other articles of equal utility to the country and commerce" (*emphasis added*); completed "elegant Botanical Drawings with their descriptions (in Number 500)" taken from nature; written a dissertation on the fever bark, a treatise on the cultivation and manufacture of indigo, another on the Hindu method of making sugar in the Circars, a third and fourth on the manufacture of raw silk and the cultivation of coffee; and completed, as well, instructions and drawings for promoting an inquiry after nutmeg and cloves.[43] And if all this were not enough, the doctor had gone above and beyond the call of duty, succoring the peasants of the neighboring *zamindaries* at a time of dearth, taking pains

> in instructing the people of the Rajahs of Pettypore (Pithapuram) and Peddypore (Peddapuram), who are his neighbours, in the art of manufacturing their sugars, by bringing people from Ganjam, where it is better understood (which he found it difficult to accomplish), to teach the process, with a liberality of disposition that does him much credit[.][44]

Now he wished to lease land in Corconda to set up an experimental farm, which was to be a site for growing vegetables and grains to sustain the poor, as well as for producing commercially useful crops. Such paternalism was common enough among Scottish advocates of improvement, but especially typical of one trained at Edinburgh's Medical School, where natural history, botany, and chemistry were taught in a manner that serviced agriculture and the related arts, and where the ideology of improvement insisted that progress in agriculture must needs depend upon those with the education and resources to follow the science underlying it, namely the landed and the learned.

Roxburgh's stewardship instincts had earlier manifested themselves in a 1791 letter to Joseph Banks in which he expressed his bitterness at the Company's hunger for profits at the expense of rational estate management:

> The Famine...which begins to rage with double force, owing to a failure of the usual rains, a continuance of such distressing misery constantly before our Eyes with/out/the power of relief, is grievous beyond description, & often makes me think of returning to my native Country...how often have I in vain urged the propriety of introducing such vegetables as would yield to the miserable poor some sort of sustenance during these times.[45]

A year later things were no better: "I have no doubt but the Company's investments may still be got up," he wrote to Robert Kyd at Calcutta, who had begun a botanic garden there to address just such concerns, "*but as they are not a mere Mercantile Body, other objects ought to be attended to,* & the queries you have put me in your last letter, convinces me that you are in Bengal possessed of some foresight."[46]

There was also a deeply personal angle to Roxburgh's paternalism, which Ross passed on to the Board. Roxburgh wished to acquire landed property in India that he could pass on to his heirs:

> Indeed he also says, that nothing will induce him to make a much longer stay in India, than being possessed of improveable Land Property, which cannot be taken from him, when it is improved; but that he may then reap the fruits of his Labour, and be certain that the benefit will devolve to his children, or to whoever else he may see it convenient to make it over.[47]

Famine was the enemy of public order, but it had occurred before in the region, so one could conjecture that if the Board now acquiesced in Roxburgh's righteous desire for distributive justice (labor, he claims, creates property) it was for other reasons entirely. The clues are in Ross's letter.

The letter is, at first glance, a recommendation from a patron supporting a client's request for accommodation from his superiors and detailing Roxburgh's services to the Company in order to push his achievements out into the (limited) public sphere of the EIC's administration in India. Any formal address to the Board of Revenue, Ross knew, would be included in the Revenue

Consultations for the year, and copies would be sent to the subordinate factories on the coast, as well as to Bengal and London.

Ross had maintained a private correspondence with Roxburgh over the years, a circumstance he described as an accident born of disinterested friendship on the part of a powerful local man for a countryman who was doing his utmost to promote the interests of "country and commerce":

> And this as well in the sentiments of friendship that subsisted between us, as in the opinion, that I should thereby be the more induced to render him from time to time, (as indeed it has been my best and unremitted endeavours to do), such assistance as it might be in my power to afford, in forwarding all his pursuits—and in rendering him those acts of kindness, in facilitating the Dispatch of his frequent intercourse with the Government here and with the Court of Directors, and his other connections in England, as my situation and constant residence, should enable me to accomplish, with less loss of time and fewer disappointments than could otherways have happened.[48]

In reality there was nothing accidental about such private communication. The members of the Board, Oakeley, White, and David Haliburton would have known exactly how to interpret Ross's rhetoric, and to assess its import for *their* private endeavors.

Private correspondence was the flip side, "the constant shadow" of the collective decisions and public communications recorded in the Consultation Books, which produced the public space of each factory. The business of recording every letter received or sent, of registering every action taken as consensual while noting dissenting positions, was meant to act as a check on the private interests of members of the Company administration. The aim was to detect the risk of, if not prevent, any one member of a Company board, usually the president or chief, from gaining the upper hand and shaping decision-making to his private ends. This was important in a context where Company resources were used equally for public and private trade, the same Indian merchants serviced both sides of the trade, and, as is evident from recent research, Company servants with greater seniority and responsibility gained the most from private trade.[49]

The streams of private correspondence that flowed alongside public communications, however, created a separate information network that worked to disrupt official channels and organize private trade and patronage relations outside Company spaces. As a free merchant, Ross's private communications were no concern of factory chiefs or Company directors in London. Not so, Roxburgh's. Roxburgh had done everything possible to direct useful information toward Ross, enabling his trading activities in exchange for the resources to carry on his researches and—importantly—his own private trade. There is evidence of Roxburgh's private connection to the President of the Board, Charles Oakeley, as well, so a letter from Ross reminding the Board of Revenue of Roxburgh's achievements and hinting at the loss to all of them should Roxburgh decide to leave India, could not but have alarmed them.

In the event Roxburgh got his farm; and it was to have an extended afterlife as the template for the Calcutta Botanic Garden, the most important scientific institution in all of British Asia. Earlier, Ross had also brokered Roxburgh's appointment as Professor of Botany and Natural History to the Company in the teeth of the Physician-General's displeasure.[50] The post had first been held by Linnaeus's student, the Dane, John Gerard Konig, and then by the Scottish naturalist and surgeon Patrick Russell. When Russell resigned in 1789, he recommended Roxburgh as his replacement. Here, however, both Ross and Roxburgh ran into trouble.

The role and place of the Company's medical service was a difficult tension at the heart of EIC state-making. The question whether it was primarily military or civil was debated throughout the latter half of the eighteenth century. Medics were lured out to India on warrants that gave them the freedom to chose their place within the service and generous scope to trade and build private fortunes. This changed once the late-eighteenth-century imperial proxy wars in the Carnatic made the deployment of surgeons with Company troops imperative. The Indian Medical Service (IMS) was twice split into two separate services, civil and military, in 1766 and 1796, but the separation was found impracticable and the service was reunited on both occasions. Various arrangements were tried out until 1788, when Governor General Lord Cornwallis drew up a long minute on the IMS, which sought to define and rationalize it by making Company physicians commissioned officers in the military.[51] Defending the Company's trade required such a step, and Cornwallis' measure sought to ensure that a militarized structure of promotions and rank would limit civil medical employment. The rub lay in curbing the activities of medics, who, as civil officers, were responsible for the natural knowledge-making that underpinned that trade.

In the Madras Presidency, promotion from Assistant Surgeon to Surgeon now meant appointment to a regiment or garrison, catching out both Roxburgh and Ross. Roxburgh was appointed to the third European regiment stationed north of Samalkota, which would have made it impossible for him to pursue either trade or natural history—unless a case could be made for the vital importance of his services as Company Naturalist at Samalkota. This Ross successfully did, forwarding a memorandum from Roxburgh to Governor Campbell and repeatedly addressing the Board of Revenue on the subject of Roxburgh's salary as Professor of Natural History. "So you have nothing to do now," he wrote Roxburgh on February 7, 1790, "but to persevere chearfully in the pursuit of your laudable Enquirys—for your own Credit, & the good of the Community & if I can any way forward your wishes by such humble means as may be in my power, I shall be well pleased."[52] He was as good as his word, and his intervention helped the institution of Company Naturalist emerge in south India. By 1794, when Roxburgh was ready to move on to Calcutta to assume charge of the Calcutta Botanic Garden, it was clear that the Company's naturalists would soon be (medical) men without other responsibilities to distract them.

Roxburgh had been rather hasty in making promises to one Dr. Wright (another Edinburgh-trained medic) that the latter could succeed him at Samalkota as botanist-naturalist. Ross, who was obviously aware that fledgling institutions were often concretized through the actions of the men first associated with them, cautioned Roxburgh not to be "more ready than is necessary...to do anything that may throw yourself...at a disadvantage by giving up or even yielding more that belongs to the line of Botanist & all its connections as well on this Coast as at Bengal." To firm up the identification of the post of Company Naturalist with his client, he engineered an opportunity for Roxburgh to be the one to draw up a plan of the responsibilities of the post and ensure that it be followed through. On March 4, 1794, Ross reported a conversation with Captain Alexander Read, who would become famous for his revenue administration of the *Baramahal*. It was Read, he told Roxburgh,

> who suggested that a person of Dr. Wright's abilities should be employed in a separate way—as you have been—& as Konig and

Russell were—& that if he could speak of it to the Board that he would do—but that it would——(word unclear) no purpose to do so, and that he thought you might without impropriety or Offence & recommend something of a Plan for such Enquirys & Researches, as might be pursued all over the great Territories of the country for Discoverys and Information into Botany Natural History and Mineralogy that may be submitted to the Court of Directors.[53]

The men in Ross's circle were to make a concerted effort with the powers-that-were: "You and I and everybody else would write also to Dr. Russell, Mr. Del—, Dr. L and you to Mr. (word unclear) & to Sir Joseph Banks which would probably [be] needed."[54] All was in train, then, for the office of Naturalist to become an essential part of the early modern colonial state. It would not be easy though, and the case would have to be made over and over again into the early decades of the nineteenth century each time a new naturalist was required to augment a state project of exploration or extraction. But the process had begun, appropriately enough, through the relationship of a private merchant keenly alive to the importance of natural knowledge to his trade, and his client, an Edinburgh medic eager to purvey such knowledge and rise above his station to take his place in the world of British savants.

* * *

There was a quid pro quo in store for Roxburgh in return for Ross's patronage, and it would prove greater than mere assistance with his trading ventures. From 1790 to 1792, the Circars, as already noted, were gripped by a terrible famine brought on by a failure of the monsoon rains. The economic life of the region was severely dislocated and Company profits took a tumble. A concerned Andrew Ross, fearful for his income, in consort with an equally worried Cornwallis, determined to assess the possibility of building canals from the Godavari and Krishna rivers to water the Circars and keep commerce going. Ross ordered Roxburgh to conduct the survey, which his client could not refuse to do because of his deep sense of obligation to his patron.

The project for "Watering the Circars," as it became known in the exchange of letters between Ross and his colleagues, suggests several things about the making of the EIC state: that building a bureaucratic state in India was a collaborative enterprise between

EIC officials (Oakeley and Cornwallis) and private merchants (Ross), that clientage relations between private men and Company men (Ross and Roxburgh) played an important role in its construction, and that patron-client ties may have served to embed the state in regional society by producing state authority as a patchwork of limited local relations. Nothing came of the venture. (It would be well into the nineteenth century before large infrastructural projects—building dams, canals, and roads—became important activities of the colonial state.) But the project provides a window to the personal and political relations through which stateliness was enacted at this early colonial moment. My claim is that lateral ties (expressed through private trade relations) played a greater role than vertical relations (between the Board of Revenue and the staff of collectorates, for example) in helping the early colonial state cohere. Roxburgh's recognition of the urgent need to improve the Circars did not mean he fell in without murmur with Ross's plans. On the contrary, he was hesitant at the prospect of upstaging his colleagues at Masulipatnam who may have wished to claim the credit and the limelight (and thus achieve entry into the "public" space of Company politics) for carrying out the survey mooted by Ross. But he was overruled impatiently by his patron, who had taken charge by cutting through Company procedure and protocol with his proposal, laying bare the reticulation of private merchant and Company interests:

> Sir Chas Oakeley has also rec^d an ans^r to his letter to Lord Cornwallis upon the Subject, who expresses the highest satisfaction at what is proposed & earnestly recommends the speedy execution of it—so that it does not seem improbable, that something will be determined upon, soon, for the trial of more expeditious means of furnishing supplies of water to those parts of the Masulipatnam & the Guntoor Districts that Maj^r Beatson alludes to, than can be obtained from the slow process of Surveys and Levels—which may be resorted to afterwards.[55]

If he wished to be nice in matters of rank and precedence he had better think again, and take a look at what was said of his Masulipatnam colleagues in Mr. White's Minute:

> That they have at no time furnished any information that has been of use to the public and after this well founded Character they will now hardly expose themselves so far to find fault in any public manner with the laudable exertions of an individual who endeavour (sic) to bring forward what will tend to the good of the Country.[56]

Charles White's minute was, in fact, an indictment of the patronage wielded by the chief of the factory at Masulipatnam and his council who enhanced their local authority (and private fortunes) by colluding with native revenue officials. And his recommendation that the Northern Circars be divided into collectorships along the lines adopted in Bengal was meant to build the state's power in the locality by separating commercial and revenue administration, and embedding Company revenue personnel, "that they might, by a residence and local knowledge with proper inquiries in the respective districts" prevent any future failure of stewardship. An oblique reference to the necessity of appointing "persons of moderation, industry, good capacity and of honourable character" because "[w]hen servants of the company holding such situations, aim at the rapid acquisition of a large fortune, many inconveniences must ensue," pointed directly to the conundrum of private trade.[57] Nevertheless, it was through a relationship nurtured through such trade that information relevant to the appointment of collectors was gathered—Ross gave Roxburgh that responsibility as well, as part of his survey of the Circars.[58]

A notable aspect of "Watering the Circars" was Ross's ability to draw information to him from Company personnel and then diffuse it in a manner that kept him in the "public" eye, while issuing a string of private commands to Roxburgh, which effectually turned the Doctor into his spy. Roxburgh's protestations that he did not wish to survey the Circars behind the backs of his Masulipatnam colleagues was countered with the remark that he need not consciously dissemble:

> In saying all this I do not mean that you should appear avowedly as employed in this way—You go into the Country in Botanical researches in pursuit of Nerium—Fever Bark—other important objects of Natural History—& in exploring these you meet with the other Subjects & examine them as every good man ought to do who has it in his power & meets with so much of public approbation as you have done.[59]

He thus connected parallel streams of information from Company sources and from Roxburgh and his local informants, much as his project sought to join the waters of the two southern rivers, the Krishna and the Godavari, flowing in opposite directions. He would later cannily direct part of his information channel above ground, releasing selected papers and correspondence to Alexander Dalrymple, who published them in the second volume of his *Oriental Repertory*—winning Roxburgh yet more "public approbation."[60]

The project itself was simple: The Northern Circars, five in number, Chicahole, Rajahmundry, Kondapalli, Ellore, and Guntur (in modern Andhra Pradesh and Odisha), composed a strip of land along the western side of the Bay of Bengal. The Eastern Ghats, a discontinuous chain of mountains cut through by four rivers including the longest river in south India, the Godavari, and the Krishna south of it, formed the western part of the Circars. Parts of the Circars lying between the two rivers were rain deficient and at the mercy of droughts and famines, so it was proposed to the Directors of the EIC that they consider building a canal system between the two rivers to direct their seasonal overflow to "water the Circars." Ross was not the first to broach such a scheme but he was the first to receive the approbation of Government. Dalrymple's narrative explained that at least two other such proposals had been brought forward, first by a Mr. John Sullivan in February 1779, and then by Lieutenant Lennon in 1788, "which, unfortunately for The Public, was not accepted."[61] The public was to be more fortunate in the case of Ross's proposal.

Both Sullivan and Lennon had evoked familiar tropes. The first was paternalistic care and profit:

> Without the aid of such assistance, the Spirit of Industry would at times be excited in vain; and the endeavours of the husbandmen instead of being rewarded with plenty, would in such event be productive of want and distress...Happily the means of providing against so dreadful a misfortune in these Provinces, are within the power of Government, and may be attained without any Considerable expence, and with the advantage of occasioning a great encrease in their cultivation and productions.[62]

The second, precise information:

> The proper Management of the Revenues of this Country can derive no greater Assistance from any thing, than good geographical plans of all the separate districts, upon a scale sufficiently large, to set clearly before the View the different kinds of Soil, and the exact quantity of cultivated ground, to ascertain the precise limits and boundaries of each division, to remark the progress of neglect, or decay, and particularly to point out the possibility of Improvement, or Cultivation.[63]

But it was Ross who carried the day.

Part of Ross's success resulted from the Governor's recognition that he was a center of calculation in the Madras Presidency *and* a point of passage for information of various kinds. The bundle of papers he forwarded to Oakeley for his consideration contained letters between Roxburgh and Colonel Robert Kyd, Dr. James Anderson, the Physician-General, and Kyd, and interestingly a letter from a sea captain and free merchant, George Baker, commenting (at Ross's request) on Roxburgh's letter to Kyd.[64] Baker, who built and ran the Madras waterworks, was described by Holden Furber as "first citizen without portfolio, consulted on every occasion of importance by government," a vital participant in Company institution building.[65] Ross's accompanying memo to Oakeley was terse, a single sentence:

> The three papers which I have now the honour of conveying into your hands, on the Subject of furnishing supplys of Water for the cultivation of the Country in the Northern Circars, being of importance too conspicuous to require any observations from me, I will only say that I consider it as my good fortune, that my intimacy with the three excellent & able men, from whom I have obtained them, gives me the opportunity of delivering them to you—which I do in the full persuasion that the very interesting designs which are there recommended, will meet with your best attention.[66]

The tone of this memo and of other communications between Ross and the Board of Revenue are remarkable for his expectation that his instructions would be followed through by the Board, an expectation seldom, if ever, belied.

Roxburgh's successes with Ross' assistance were many. By the time he left India, he had earned a substantial fortune of 50,000 pounds, no mean feat for an EIC surgeon whose pay averaged 30 pounds a month.[67] Furthermore, his unassailable stature as a naturalist earned him entry into Britain's premier scientific circles as well as two gold medals from the Society for the Arts for his "valuable Communications on East India products." When he died in 1815, he had firmly set his son on a path into the gentry and secured his own claim to his posthumous title as "The Father of Indian Botany." A title, suitably enough, thought up and popularized by his son.

* * *

This chapter has explored how an eighteenth-century medic with an Edinburgh education could make his way in the IOW. I have argued that studying medicine at Edinburgh equipped men like Roxburgh with the skills they needed to earn large fortunes for themselves through trade, even as they made the natural knowledge that underpinned EIC state building. Roxburgh's *Nerium* had the potential to alter the natural environment in ways that promoted governance (it was logistical knowledge) but it was, first of all, an object of commerce (Figure 5.1).

A medic-on-the-make was a familiar enough sight in EIC India to be satirized in the *Calcutta Review* of 1854 as the "Philosophic Surgeon"

Figure 5.1 Herbarium sheet, *Nerium tinctorium*, labelled in Roxburgh's hand (courtesy: Henry J. Noltie, Royal Botanic Garden, Edinburgh).

who, on his way to his indigo factory, would inquire of the native doctor—"Any thing today"—and, upon receiving the ready answer, "All's well, Lord of the World! only five men dead," would exclaim cheerfully—"good, very good"—and canter gaily about his business.[68]

The commercial being of the EIC state enabled the development of this social type, even as it allowed the growth of a form of shared rule in Madras that accorded significant authority to private merchants in Company settlements. Such merchants could have a critical say in the development of the state's institutions, as I have shown in my discussion of the politics surrounding the formation of the post of Company Naturalist.

The social-economic relations that underpinned the Company state were produced through familial ideology operating in different registers. By accepting Ross as his patron, Roxburgh was able to realize his commitment to stewarding the Circars. By accepting Ross as his patron, Roxburgh was able to realize his commitment to stewarding the Circars (the improving imagination was rooted in paternalism), as well as his dream of accumulating property that would devolve to his children. The two were interlaced, so that we could gloss Julia Adams's observations on the patrimonial nexus of seventeenth-century state making to note that the *affect* produced by the convergence of paternal authority and patriarchal status in patrimonialism continued well into the late eighteenth and nineteenth centuries and shaped the actions of men who were part of the process of colonial-state-institution building, such as Ross and Roxburgh. This was sharply manifested in a dispute over the succession to the Superintendentcy of the Calcutta Botanic Garden, which broke out when Roxburgh was ready to retire from the post.[69]

The emergence of the bureaucratic EIC state in India did not exclude the working of familial ideology in ordering relations between male bureaucrats, and between them and those placed in their control through the contingency of history.[70] Nor did it separate all state functionaries from the networks of exchange established through private trade. Instead, the rationality of the early colonial state subsumed these features within its forms of governance. It was one way in which the *Company*-State organized its stateliness.

Acknowledgments

I thank Anna Elizabeth Winterbottom for her comments and careful editing of this chapter; K. S. Subramanian for his help with the Tamil; Mr. Balasubramaniam and Mrs. Raje at the Sarasvati Mahal Library for assistance with the Modi, and Chandra Mukerji, Robert Eric Frykenberg, and Santhi Hejeebu for useful suggestions on an earlier version of the chapter. The usual disclaimers apply.

Notes

1. Philip J. Stern (2011), "Introduction," *The Company-State: Corporate Sovereignty & the Early Modern Foundations of the British Empire in India* (New York, NY: Oxford University Press), pp. 3–15.
2. Ibid.
3. See Holden Furber (1951), *John Company at Work* (Cambridge: Harvard University Press), pp. 192–207; and the essay by Holden Furber (1997), "Madras in 1787" in *Private Fortunes and Company Profits in the India Trade in the 18th Century*, ed. Rosane Rocher (Brookfield, VT: Variorum).
4. The argument here distinguishes state making in Madras from accounts of the conflicts caused by the Company's intermingled commercial and political roles. For a contemporary narrative, see Harry Verelst's complaints about the shortage of currency in Bengal after the acquisition of *Diwani*, which strangled Bengal's commerce. Harry Verelst (1772), *A View of the Rise, Progress, and Present State of the English Government in Bengal* (London: J. Nourse), pp. 84–104. For a discussion of the issue, see Robert Travers (2007), *Ideology and Empire in Eighteenth-Century India* (New York, NY: Cambridge University Press), p. 80.
5. Julia Adams (2005), *The Familial State: Ruling Families and Merchant Capitalism in Early Modern Europe* (Ithaca, NY: Cornell University Press), p. 31.
6. I use the term "medic" rather than the contemporary denomination "surgeon" to signal how the careers of men like Roxburgh exceeded conventional functional boundaries.
7. John Walker (c. 1780), "Lectures on Natural History," *Dc. 217–21*, 5 vols., Centre for Research Collections/Edinburgh University Library; for a discussion, see Minakshi Menon (2013), *Making Useful Knowledge: British Naturalists in Colonial India, 1784–1820*, PhD diss., History (Science Studies), University of California, San Diego (Ann Arbor: ProQuest), Chapter 3.
8. Walker, "Lectures", Vol. 3, *Dc. 219*, f. 7.
9. J. V. Golinski (1988), "Utility and Audience in Eighteenth-Century Chemistry: Case Studies of William Cullen and Joseph Priestley," *British Journal for the History of Science*, 21: 1–31.
10. See Steven Shapin (1974), "The Audience for Science in Eighteenth Century Edinburgh," *History of Science*, 12: 95–121.

11. Harold J. Cook (2007), *Matters of Exchange: Commerce, Medicine, and Science in the Dutch Golden Age* (New Haven, CT, and London: Yale University Press), p. 411.
12. Ibid., pp. 267–303.
13. Ibid., p. 411.
14. A. L. Donovan (1975), *Philosophical Chemistry in the Scottish Enlightenment* (Edinburgh: Edinburgh University Press); and discussion of "experimental history" in Ursula Klein and Wolfgang Lefèvre (2007), *Materials in Eighteenth-Century Science: A Historical Ontology* (Cambridge, MA, and London: MIT Press), pp. 22–28.
15. George K. McGilvary (2008), *East India Patronage and the British State: The Scottish Elite and Politics in the Eighteenth Century* (London and New York, NY: Tauris Academic Studies), pp. 113–114.
16. Arjun Appadurai (1986), "Introduction: Commodities and the Politics of Value," in *The Social Life of Things: Commodities in Cultural Perspective*, ed. Arjun Appadurai (Cambridge and New York, NY: Cambridge University Press), pp. 41–43.
17. See Dirom G. Crawford (1914), *A History of the Indian Medical Service 1600–1913*, 2 vols. (London: W. Thacker), Vol. 1, pp. 253–260 and 292.
18. Chandra Mukerji (2010), "The Territorial State as a Figured World of Power: Strategics, Logistics, and Impersonal Rule," *Sociological Theory*, 28 (4): 402–424: 404.
19. Ibid.; and Chandra Mukerji (2009), *Impossible Engineering: Technology and Territoriality on the Canal du Midi* (Princeton, NJ, and Oxford: Princeton University Press).
20. The Northern Circars (where Roxburgh was stationed) were a particularly propitious site for the generation of private fortunes for reasons noted by Furber, *John Company*, pp. 198–199.
21. Shapin, "Audience," 102.
22. The term is Simon Schaffer's. See his (2003), "Enlightenment Brought Down to Earth," *History of Science* 41: 260. Schaffer uses "pattern science" to refer to the redefinition of agronomy in the eighteenth century. The meaning of the word changed from being a written account for purposes of rural administration to experimental-philosophical enquiry into the management of agricultural production.
23. Andrew Mackillop (2003), "Fashioning a 'British' Empire: Sir Archibald Campbell of Inverneil & Madras, 1785–9," in *Military Governors and Imperial Frontiers c. 1600–1800: A Study of Scotland and Empires*, ed. A. Mackillop and Steve Murdoch (Leiden and Boston, MA: Brill), pp. 205–231.
24. Thomas Ertman (1999), "Explaining Variation in Early Modern State Structure," in *Rethinking Leviathan*, ed. John Brewer and Eckhart Hellmuth (London: German Historical Institute), p. 49.
25. John Gascoigne (1998), *Science in the Service of Empire: Joseph Banks, the British State and the Uses of Science in the Age of Revolution* (Cambridge: Cambridge University Press), pp. 111–146.

26. Miles Ogborn (2007), *Indian Ink: Script and Print in the Making of the English East India Company, 2007* (Chicago, IL: University of Chicago Press).
27. Alexander Dalrymple (1794), *Oriental Repertory Published at the Charge of the East India Company*, 4 vols. (London: Printed by George Biggs, etc.), Vol. 2.
28. Pa. Subramanian, ed. (1989) *tañcai marāṭṭiya maṉṉar mōṭi āvaṇat tamiḻakamuṉ kurippuraiyum mutal tokuti (Thanjāvur Maratha Kings' Modi Documents of Tamilakam and Notes*, Vol. 1) (Thanjavur: Tamil University), pp. 622–631 (Modi Bundle No. 43C, Thanjavur Maharaja Serfoji's Sarasvati Mahal Library).
29. Andrew Ross to William Roxburgh, June 2, 1788; December 19, 1788; December 23, 1788, *Roxburgh Correspondence/Natural History Museum*, London (hereafter RC/NHM).
30. Ross to Roxburgh, Fort St. George, June 2, 1788, RC/NHM, f. 156.
31. Ross to Roxburgh, Fort St. George, December 23, 1788, RC/NHM, f. 151.
32. Ross to Roxburgh, December 23, 1788.
33. The nature of their partnership may have changed with the Cornwallis Code of 1793, in which the right to trade privately was restricted to the EIC's commercial service (though EIC medics seem to have had the right to trade privately even later). One route taken by Company personnel was to direct their capital for investment through agency houses, which became the unit of British private trade in India. There is frequent mention of Roxburgh's agent, James Amos, the founder of Amos & Bowden, later Michell, Amos & Bowden, in his correspondence. See Furber, *John Company*, for Amos, p. 114. On the agency houses, see Amales Tripathi (1956), *Trade and Finance in the Bengal Presidency 1793–1833* (Bombay: Orient Longman); S. B. Singh (1966), *European Agency Houses in Bengal (1783–1833)* (Calcutta: Firma K. L. Mukhopadhyay); and more recently, Tony Webster (2005), "An Early Global Business in a Colonial Context: The Strategies, Management and Failure of John Palmer and Company of Calcutta, 1780–1830," *Enterprise & Society*, 6 (1): 98–133, who points out that even before the Cornwallis reforms free merchants had begun to pool their resources as their ventures required large amounts of capital to turn a profit. Roxburgh may well have begun to make his savings over to Ross before 1793, and thus been unaffected by the changes introduced after that date.
34. Dalrymple, *Oriental Repertory*, 1791, Vol. 1, no. 1, pp. 1–44.
35. Ibid., p. 42.
36. Ibid., p. 44.
37. W. Roxburgh and Alexander Anderson (1810), "Papers in Colonies and Trade," *Transactions of the Society, Instituted at London, for the Encouragement of Arts, Manufactures and Commerce* 28: 249–316, at 270.
38. For the local aspects of building the colonial state's bureaucracy in the Circars, see Robert Eric Frykenberg (1965), *Guntur District 1788–1848: A History of Local Influence and Central Authority in South India* (Oxford: Clarendon Press). For a critique, see Bhavani Raman (2012), *Document Raj: Writing and Scribes in Early Colonial South India* (Chicago, IL, and London: University of Chicago Press), pp. 24–25.

39. Walter Kelly Firminger, ed. (1969 [1918]), "Appendix No. 14 MINUTES of Mr. C. N. White (Member of the Board of Revenue at Fort St. George); dated 14th February, and 25th March, 1793" *The Fifth Report from the Select Committee of the House of Commons on the Affairs of the East India Company, 28th July, 1812*, 3 vols. (New York, NY: Augustus M. Kelley), Vol. 3, p. 122.
40. Walter Hamilton (1820), "The Northern Circars," *A Geographical, Statistical and Historical Description of Hindostan and the Adjacent Countries in Two Volumes* (London: John Murray), Vol. 2, pp. 63–65. A comprehensive discussion of the *pālaiyakkārar* can be found in Nicholas B. Dirks (1987), *The Hollow Crown: Ethnohistory of an Indian Kingdom* (Cambridge and New York, NY: Cambridge University Press). Dirks's work is specifically concerned with one "little kingdom," *Pudukkottai*, in present-day Tamil Nadu, but his observations on these old regime chieftains can be said to apply to those of the Andhra country as well; also see pp. 19–35, *Fifth Report*, Vol. 2, p. 4; and Charles Wilkins's Glossary in *Fifth Report*, Vol. 3, p. 41, which notes that "*Polligar[s]*" [were] military chieftain[s] in the peninsula similar to the hill *Zemindar* in the Northern *Circars*[.]"
41. *Proceedings, Revenue Department* 1793, Vol. No. 53 B, p. 2338, Tamil Nadu State Archives, Chennai.
42. Andrew Ross to David Haliburton Esq Acting President and Members of the Board of Revenue, June 20, 1793, RC/NHM (large envelope).
43. *Roxburgh Manuscripts*, MSS/EUR D809, Asia Pacific and Africa Collections, British Library (hereafter APAC/BL).
44. Andrew Ross to BOR, June 20, 1793 (emphasis in original).
45. Roxburgh to Banks, Samalkota, August 30, 1791, in Neil Chambers, ed. (2010), *The Indian and Pacific Correspondence of Sir Joseph Banks*, 6 vols. (London: Pickering and Chatto), Vol. 3: Letters 1789–1792, Letter [194], p. 278.
46. Extract of a letter from Dr. Roxburgh to Col. Kyd, October 17, 1792, MSS/EUR D809/APAC/BL (n. p.) (emphasis added).
47. Ross to BOR, June 1793.
48. Ibid.
49. Ogborn, *Indian Ink*, Chapter Three; Santhi Hejeebu (2005), "Contract Enforcement in the English East India Company," *The Journal of Economic History*, 65 (2): 496–523: 511, Figure 2: "Remittances Increase With Seniority, 1746–1756"; and on p. 512, Table 1: "Remittances Rise Over the Course of a Career" (Hejeebu's figures hold for the period up to 1757). There is no research for the later eighteenth century, but Furber hints that conditions were similar in the 1780s and 1790s.
50. Ross to Roxburgh, May 13, 1791, RC/NHM.
51. The minute is reproduced in full in Crawford, *A History*, Vol. 1, pp. 253–260.
52. Ross to Roxburgh, Madras, February 7, 1790, RC/NHM.
53. Ross to Roxburgh, Madras, March 4, 1794, RC/NHM, f. 58 (verso). Interestingly, Read was the maternal uncle of Helena, Lady Oakeley, and Captain Alexander Beatson.
54. Ross to Roxburgh, March 4, 1794.

55. Ross to Roxburgh, Madras, January 13, 1793, RC/NHM, f. 124 (verso) and f. 125 (recto). Beatson was Oakeley's brother-in-law.
56. Ross to Roxburgh, Madras, March 16, 1793, RC/NHM, f. 126 (verso).
57. "Appendix No. 14: MINUTES of Mr. C. N. White (Member of the Board of Revenue at Fort St. George); dated 14th February, and 25th March, 1793," in *Fifth Report*, ed. Firminger, Vol. 3, pp. 118–125; quotes on pp. 121 and 122.
58. Ross to Roxburgh, March 6, 1793, RC/NHM, f. 114 (recto).
59. Ross to Roxburgh, Madras, March 16, 1793, RC/NHM, f. 110 (verso) and f. 111 (recto).
60. Dalrymple, *Oriental Repertory*, Vol. 2, No. 1, pp. 33–84.
61. Ibid., pp. 57–66.
62. Ibid., p. 61.
63. Ibid., p. 58.
64. Extract of a letter from Dr. Anderson to Col. Kyd dated August 1792; extract of a letter from Dr. Roxburgh to Col: Kyd October 17, 1792; Mr. Baker to Mr. Ross, St. Thome, November 13, 1792; Alexander Beatson to Sir Charles Oakeley, Fort St. George, December 1, 1792; Memorandums of Major Beatson, December 5, 1792; Dr. Alexander Anderson to Mr. Ross, December 4, 1792, in MSS/EUR D809/APAC/BL (n. p.).
65. Furber, "Madras in 1787," p. 264.
66. "From Mr. Ross to Sir Chas Oakeley, Fort St. George 16th Novr 92," MSS/EUR D809/APAC/BL (n. p.).
67. It could be argued that while in Madras, Roxburgh's income was augmented by a variety of perks such as a generous house allowance, and so on. I find no evidence for such Company-granted largesse (indeed his pay as Company Naturalist was the subject of extended negotiation, as described above). Apart from his private trade, Roxburgh was entrusted with the right to collect land duties in Corcondah District. *Proceedings, Revenue Department,* 1793, TNSA, pp. 2338–2339.
68. "ART. VIII—Alphabetical List of the Medical Officers of the Indian Army; with the dates of their respective Appointments, Promotion, Retirement, Resignation, or Death, whether in India or Europe; from the year 1764, to the year 1838. Compiled by Messrs. Dodwell and Miles." *Calcutta Review,* 23 (45) (July 1854): 217–254, at p. 242.
69. Adams, *Familial State,* p. 29; and Menon, *Making Useful Knowledge,* pp. 286–288, for Roxburgh's efforts to replace himself with his son at the Calcutta Botanic Garden.
70. Bhavani Raman has discussed how family and patronage relations were important elements in the structuring of the EIC's *kacceri* (district office) in early colonial Madras. She points out that Company officials reified and altered preexisting familial networks, making the collector's patronage the decisive element in *kacceri* appointments. Raman shows similar imperatives (information collection) at work in the lower echelons of the colonial bureaucracy that I describe among Company elites. Bhavani Raman (2008), "The Familial World of the Company's *kacceri* in Early Colonial Madras," *Journal of Colonialism and Colonial History,* 9: 2. Also see her *Document Raj,* pp. 29–43.

Bibliography

Abas, S. J. and A. S. Salman (1995). *Symmetries of Islamic Geometrical Patterns*. Singapore and New Jersey: World Scientific.
Abu-Rabia, A. (2005). "The Evil Eye and Cultural Beliefs among the Bedouin Tribes of the Negev, Middle East [1]." *Folklore*, 116 (3): 241–254.
Adams, J. (2005). *The Familial State: Ruling Families and Merchant Capitalism in Early Modern Europe*. Ithaca, NY: Cornell University Press.
Agırakça, A. (2004). *İslâm Tıp Tarihi Başlangıçtan Vii./xiii. Yüzyıla Kadar*. İstanbul: Çağdaş Basın Yayın Ltd. Şti.
Agırakça, A. (2010). *İslâm Tıp Tarihi (History of Islamic Medicine)*. İstanbul: Akdem Yayinlari.
Ahmad F., N. Tabassum, S. Rasool (2012). "Medicinal Uses and Phytoconstituents of *Paeonia officinalis*." *International Research Journal of Pharmacy*, 3 (4): 85–87.
Ahmed, S., S. Bamofleh, and M. Munshi. (1989). "Cultivation of Neem (*Azadirachta indica*, Meliaceae) in Saudi Arabia." *Economic Botany*, 43(1): 35–38.
Ainslie, W. (1813). *Materia Medica of Hindoostan, and Artisan's and Agriculturist's Nomenclature*. Madras: Government Press.
Ainslie, W. (1826). *Materia Indica; Or, Some Account of Those Articles which are Employed by the Hindoos and Other Eastern Nations, in Their Medicine, Arts, and Agriculture*. London: Longman, Rees, Orme, Brown, and Green.
Alam, M. (2004). *The Languages of Political Islam: India, 1200–1800*. Chicago, IL: University of Chicago Press.
Alavi, S. (2008). *Islam and Healing: Loss and Recovery of an Indo-Muslim Medical Tradition, 1600–1900*. New York, NY: Palgrave Macmillan.
Aldridge, R. D. and D. M. Sever (2011). *Reproductive Biology and Phylogeny of Snakes*. Enfield, NH, and Boca Raton, FL: Science; marketed and distributed by CRC Press.
al-Ghazal, S. K. (2007). *The Valuable Contributions of Al-Rāzī (Rhazes) in the History of Pharmacy*. Manchester: Foundation for Science Technology and Civilization.
al-Ḥamawī, Y. i. A. (1866). *Muʻjam Al-Buldān (Geographical Dictionary)*, edited by F. Wüstenfeld. Leipzig: F.A. Brockhaus.

al-Maqdisi, M. b. T. (1899–1919). *Al-Bad' Wa-'t-Ta'rīḫ*, 6 vols., edited by C. Huart. Paris.
al-Masʿūdī, ʿ. I. a. (1965). *Kitāb at-Tanbīh Wa-Ăl-Išrāf*. Beirut: Khayyat.
al-Qifṭī, D. a. A. l. (1903). *Ta'Rīkh Al-Ḥukamā': Wa-Huwa Mukhtaṣar Al-Zwzanī Al-Musammā Bi l-Muntakhabāt Al-Multaqaṭāt Min Kitāb Ikhbār Al-ʿUlamā' Bi-Akhbār Al-Ḥukamā' (History of Physicians...)*, edited by J. Lippert. Leipzig.
al-Qifṭī, D. a. A. l. (1908). *Ekhbar Alolama' be Akhbar Alhukama (Informing Scholars of the Stories of Philosophers)*. Egypt: Dar al-Saadeh.
al-Rāzī, A. B. M. i. Z. (1978). *Mā Alfark? (What's the Difference?)*, edited by S. Katayeh. Aleppo, Syria: University of Aleppo.
al-Rāzī, A. B. M. i. Z. (1987). *Al-Manṣūrī fī Al-Ṭibb (The Medical Book of Mansur)*, edited by H. a.-B. Ṣiddīqī. al-Kuwayt (Kuwait): Maʿhad al-Makhṭūṭāt al-ʿArabīyah, al-Munaẓẓamah al-ʿArabīyah lil-Tarbiyah wa-al-Thaqāfah wa-al-ʿUlūm (Publications of the Institute of Arabic Manuscripts, Kuwait).
al-Rāzī, M. i. Z. (1992). *Taqāsīm Al-ʿilal (The Divisions of Diseases): Kitāb at-Taqsīm Wa-t-Tašġīr*. Aleppo: Maʿhad at-turāṯ al-ʿilmī al-ʿarabī.
al-Rāzī, A. B. M. i. Z. (2009). *Al-Tibb Al-Mulūkī (The Royal Medicine)*, edited by M. Y. Zakkūr. Jiddah: Dār al-Minhāj lil-Nashr wa-al-Tawzīʿ.
al-Rāzī, A. B. M. i. Z. and Y. Zaydān (2003). *Treatise on Gout = Traîté Sur La Goutte = Abhandlung Zur Gicht = Maqāla fī Al-Naqras*. Alexandria, Egypt: Bibliotheca Alexandrina.
al-Rāzī, M. I. a. (1962). *Kitāb Al-Hāwī fī ṭ-Ṭibb Al-Ǧuz' 11 Al-Ǧuz' 11*. Ḥaidarābād ad-Dakkān: Dā'irat al-maʿārif al-ʿutmānīya.
al-Ṭabarī, ʿA. i. S. R. (1928). *Firdaus Al-Ḥikma Fi 't-Ṭibb (Paradise of Wisdom in Medicine)*, edited by M. Z. Siddiqi. Berlin-Charlottenburg: Buch- u. Kunstdruckerei "Sonne".
al-Khaliq, A. a. (1957). *Muhammad Ibn Abi Bakr Ibn Qayyim Al-Jawziyya, al Tibb Al-Nabawi*. Beirut: Dar al-Kutub al Ilmiyya.
Anon (2000 [1781]). *Rifayee Mala*. Calicut: Thirurangadi Book Stall.
Antony, P. (2000). *PayyannurPattu*, edited by S. Zacharia (Tuebingen University Library Malayalam Manuscript Series ed.). Kottayam: DC Books.
Appadurai, A., ed. (1986). *The Social Life of Things: Commodities in Cultural Perspective*. Cambridge: Cambridge University Press.
Appiah, A. (2006). *Cosmopolitanism: Ethics in a World of Strangers*. New York, NY: W. W. Norton.
Arafath, P. K. Y. (2012). "History of Medicine and Hygiene in Medieval Kerala: 14–16 Centuries." Unpublished PhD thesis, University of Hyderabad, Hyderabad.
Arasaratnam, S. (1994). *Maritime India in the Seventeenth Century*. Delhi and New York, NY: Oxford University Press.
Arnold, D. (1988). *Imperial Medicine and Indigenous Societies*. Manchester, NY and New York, NY: Manchester University Press; distributed exclusively in the USA and Canada by St. Martin's Press.
Arnold, D. (1991). "The Indian Ocean as a Disease Zone, 1500–1950." *South Asia: Journal of South Asian Studies*, 14 (2): 1–21.

Bibliography

Arnold, D. (1994). "Colonial Medicine in Transition: Medical Research in India, 1910–47." *South Asia Research*, 14 (1): 10–35.
Arnold, D. (1996). *Warm Climates and Western Medicine: The Emergence of Tropical Medicine, 1500–1900*. Amsterdam and Atlanta, GA: Rodopi.
Attewell, G. N. A. (2007). *Refiguring Unani Tibb : Plural Healing in Late Colonial India*. Hyderabad, India: Orient Longman.
Balfour, E. (1885). *The Cyclopædia of India and of Eastern and Southern Asia: Commercial, Industrial and Scientific Products of the Mineral, Vegetable, and Animal Kingdoms, Useful Arts and Manufactures*. London: B. Quaritch.
Barbosa, D. (2009 [1865]). *A Description of the Coasts of East Africa and Malabar in the Beginning of the Sixteenth Century*, edited by H. E. J. Stanley. London: Hakulyt Society.
Bari, A. and A. Hussain (2001). "Hakim 'Imād Al-Dīn Mahmūd b. Fahkr Al-Dīn Muhammad Shirazi." *Studies in History of Medicine and Science*, 17 (1–2): 73–85.
Barnes, J., L. Anderson, and J. D. Phillipson (2007). *Herbal Medicines*. London and Grayslake, IL: Pharmaceutical Press.
Barnhill, D. L. and R. S. Gottlieb (2001). *Deep Ecology and World Religions: New Essays on Sacred Grounds*. Albany: State University of New York Press.
Bartolomeo, F. P. (1800). *A Voyage to the East Indies...* London: Printed by J. Davis; sold by Vernor and Hood; and J. Cuthell.
Bastos, C. (2005). "Race, Medicine and the Late Portuguese Empire: The Role of Goan Colonial Physicians." *Journal of Romance Studies*, 5 (1): 23–35.
Bastos, C. (2008). "Migrants, Settlers and Colonists: The Biopolitics of Displaced Bodies." *International Migration*, 46 (5): 27–54.
Bayley, E. C. (1883). "Art. I—On the Genealogy of Modern Numerals. Part II. Simplification of the Ancient Indian Numeration." *Journal of the Royal Asiatic Society of Great Britain & Ireland*, 15 (1): 1–72.
Bayly, S. (1989). *Saints, Goddesses, and Kings: Muslims and Christians in South Indian Society, 1700–1900*. Cambridge and New York, NY: Cambridge University Press.
BeDuhn, J., ed. (2009). *New Light on Manichaeism: Papers from the Sixth International Congress on Manichaeism*. Leiden and Boston, MA: Brill.
Bhaskaranuni, P. (2000). *Pathompatham Nootandile Keralam*. Trissur: Kerala Sahitya Akademi.
Binyamīn, Asher A. (1840). *The Itinerary of Rabbi Benjamin of Tudela*. London: Asher.
Bīrūnī, M. i. A. and P. Kraus (1936). *Risālah Lil-Bīrūnī fī Fihrist Kutub Muḥammad Ibn Zakarīyā' Al-Rāzī*. Bārīs: Maṭbaʻat al-Qalam.
Boag, W. (1799). "On the Poison of Serpents." *Asiatic Researches Comprising History and Antiquities, the Arts, Sciences and Literature of Asia*, 6: 112–113.
Bodeker, G., G. Burford, J. R. Chamberlain, and K. K. S. Bhat (2001). "The Underexploited Medicinal Potential of *Azadirachta indica* A. Juss. (Meliaceae) and *Acacia nilotica* (L.) Willd. Ex Del. (Leguminosae) in Sub-Saharan Africa: A Case for a Review of Priorities." *International Forestry Review*, 3: 285–298.

Bose, S. (2006). *A Hundred Horizons: The Indian Ocean in the Age of Global Empire*. Cambridge, MA: Harvard University Press.

Braudel, F. (1972). *The Mediterranean and the Mediterranean World in the Age of Philip II*. New York, NY: Harper & Row.

Brice, W. C. (1981). *An Historical Atlas of Islam*. Leiden: Brill.

Brokelman, K. (1977). *Tarikh Al-Adab Al-'Arabi (History of Arab Literature)*, edited by A. H. al-Najjar. Cairo: Dar al-Maaref.

Caldwell, S. (2001). *Oh Terrifying Mother: Sexuality, Violence and Worship of the Goddess Kālī*. Delhi and Oxford: Oxford University Press.

Cammann, S. V. R. and C. F. Bieber (1962). *Substance and Symbol in Chinese Toggles; Chinese Belt Toggles from the C.F. Bieber Collection*. Philadelphia: University of Pennsylvania Press.

Chakrabarti, P. (2006). "Neither of Meate nor Drinke, but What the Doctor Alloweth: Medicine amidst War and Commerce in Eighteenth-Century Madras." *Bulletin of the History of Medicine*, 80 (1): 1–38.

Chakrabarti, P. (2010). *Materials and Medicine: Trade, Conquest, and Therapeutics in the Eighteenth Century*. Manchester: Manchester University Press.

Chambers, N. (2010). *The Indian and Pacific Correspondence of Sir Joseph Banks, 1768–1820, Vol. 3*. London: Pickering & Chatto.

Charlier, R. H. (2009). "The Healing Sea: A Sustainable Coastal Ocean Resource: Thalassotherapy." *Journal of Coastal Research*, 25 (4): 838–856.

Chaudhuri, K. N. (1985). *Trade and Civilisation in the Indian Ocean: An Economic History from the Rise of Islam to 1750*. Cambridge: Cambridge University Press.

Chen, Yinghui, Xue, Qishan, Zheng, Peikai, and Museu de Macau, (2012). *Hai Shang Ci Lu: Yue Gang Ao Wen Wu Da Zhan = A Rota marítima Da Porcelana: Relíquias Dos Museus De Guangdong, Hong Kong e Macau = Maritime Porcelain Road: Relics from Guangdong, Hong Kong, Macao Museums*. Aomen: Aomen bo wu guan.

Chipman, L. (2010). *The World of Pharmacy and Pharmacists in Mamlūk Cairo*. Leiden and Boston, MA: Brill.

Chudan, V. T. I. (1969). *The Secret Chamber*. Trichur: Cochin Dewaswom Board.

Catlin-Jairazbhoy, A. and E. A. Alpers (2004). *Sidis and Scholars: Essays on African Indians*. Noida, UP: Rainbow.

Cochran, S. (2006). *Chinese Medicine Men: Consumer Culture in China and Southeast Asia*. Cambridge, MA: Harvard University Press.

Cook, H. J. (2007). *Matters of Exchange: Commerce, Medicine, and Science in the Dutch Golden Age*. New Haven, CT: Yale University Press.

Craig, S. R. (2012). *Healing Elements: Efficacy and the Social Ecologies of Tibetan Medicine*. Berkeley: University of California Press.

Crawford, D. G. (1914). *A History of the Indian Medical Service 1600–1913*. London: Thacker.

Critchlow, K. (1976). *Islamic Patterns: An Analytical and Cosmological Approach*. New York, NY: Schocken Books.

Dale, S. F. and M. G. Menon (1978). "Nerccas: Saint-Martyr Worship among the Muslims of Kerala." *Bulletin of the School of Oriental and African Studies*, 41 (3): 523–538.

Dalrymple, A. (1793–1797). *Oriental Repertory Published at the Charge of the East India Company*, 4 vols. London: George Biggs.

Dalrymple, W. (2009). *Nine Lives: In Search of the Sacred in Modern India*. London and Berlin: Bloomsbury.

Das Gupta, A. and M. N. Pearson (1987). *India and the Indian Ocean, 1500–1800*. Calcutta and New York, NY: Oxford University Press.

De Michelis, E. (2005). *A History of Modern Yoga Patañjali and Western Esotericism*. London: Continuum.

Desai, Z.-D. A. (1989). *A Topographical List of Arabic, Persian, and Urdu Inscriptions of South India*. New Delhi: Indian Council of Historical Research, Northern Book Centre.

Digby, A., W. Ernst, and P. B. Muhkarji (2010). *Crossing Colonial Historiographies: Histories of Colonial and Indigenous Medicines in Transnational Perspective*. Newcastle: Cambridge Scholars.

Dirks, Nicholas B. (1987). *The Hollow Crown: Ethnohistory of an Indian Kingdom*. Cambridge and New York, NY: Cambridge University Press.

Dodwell, E. and J. S. Miles (1854). "Alphabetical List of the Medical Officers of the Indian Army with the Dates of their Respective Appointment, Promotion, Retirement, Resignation, or Death, Whether in India or in Europe; from the Year 1764, to the Year 1838." *Calcutta Review*, 23 (45): 217–254.

Donovan, A. L. (1975). *Philosophical Chemistry in the Scottish Enlightenment: The Doctrines and Discoveries of William Cullen and Joseph Black*. Edinburgh: Edinburgh University Press.

Du Choisel, C. (1756). *Nouvelle méthode sure, courte et facile pour le traitement des personnes attaquées de la rage*. Paris: Chez H. L. Guerin & L. F. Delatour.

Dundes, A. (1992). *The Evil Eye: A Casebook*. Madison: University of Wisconsin Press.

Eberhard, Wolfram (1986). *A Dictionary of Chinese Symbols: Hidden Symbols in Chinese Life and Thought*. London and New York, NY: Routledge & Kegan Paul.

Elgood, C. (1951). *A Medical History of Persia and the Eastern Caliphate from the Earliest Times until the Year A.D. 1932*. Cambridge: Cambridge University Press.

Elworthy, F. T. (1895). *The Evil Eye: An Account of This Ancient & Widespread Superstition*. London: John Murray.

Ernst, W. and P. B. Mukharji (2009). "From History of Colonial Medicine to Plural Medicine in a Global Perspective: Recent Works on History of Medicine in Colonial/Postcolonial Contexts." *NTM Journal of the History of Science, Technology and Medicine*, 17 (4): 447–458.

Fabrega, H. (2009). *History of Mental Illness in India: A Cultural Psychiatry Retrospective*. Delhi: Motilal Banarsidass.

Fenger, J. F. (1863). *History of the Tranquebar Mission Worked Out from the Original Papers*. Tranquebar: Evangelical Lutheran Mission Press.

Fjelstad, K. and N. T. Hien (2011). *Spirits without Borders: Vietnamese Spirit Mediums in a Transnational Age*. New York, NY: Palgrave Macmillan.
Flood, G. (2003). *The Blackwell Companion to Hinduism*. Oxford: Blackwell.
Flores, J. M. C. d. S. (2007). *Re-Exploring the Links: History and Constructed Histories between Portugal and Sri Lanka*. Wiesbaden: Harrassowitz.
Freeman, J. R. (1999). "Gods, Groves and the Culture of Nature in Kerala." *Modern Asian Studies*, 33 (2): 257–302.
Frey, E. F. (1979). "Saints in Medical History." *Clio Medica (Amsterdam, Netherlands)*, 14 (1): 35–70.
Frykenberg, R. E. (1965). *Guntur District, 1788–1848: A History of Local Influence and Central Authority in South India*. Oxford: Clarendon Press.
Fuller, C. J. (1975). "The Internal Structure of the Nayar Caste." *Journal of Anthropological Research*, 31: 283–312.
Furber, H. (1948). *John Company at Work: A Study of European Expansion in India in the Late Eighteenth Century*. Cambridge: Harvard University Press.
Furber, H. (1997). *Private Fortunes and Company Profits in the India Trade in the 18th Century*. Brookfield, VT: Variorum.
Furber, H., S. Arasaratnam, and K. McPherson (2004). *Maritime India*. New Delhi and Oxford: Oxford University Press.
Furth, C. (1999). *A Flourishing Yin Gender in China's Medical History, 960–1665*. Berkeley: University of California Press.
Gaṇēś, K. E., ed. (2004). *Culture and Modernity: Historical Explorations*. Calicut: University of Calicut.
Gascoigne, J. (1998). *Science in the Service of Empire: Joseph Banks, the British State and the Uses of Science in the Age of Revolution*. Cambridge, UK, and New York, NY: Cambridge University Press.
Geddes, Michael (1694). *The History of the Church of Malabar...* London: Printed for Sam. Smith and Benj. Walford.
Germann, W. (1880). "Ziegenbalg's Bibliotheca Malabarica." *Missionsnachrichten Der Ostindischen Missionsanstalt Zu Halle*, 32 (3–4): 91.
Goble, A. E. (2011). *Confluences of Medicine in Medieval Japan Buddhist Healing, Chinese Knowledge, Islamic Formulas, and Wounds of War*. Honolulu: University of Hawai'i Press.
Golinski, J. V. (1988). "Utility and Audience in Eighteenth-Century Chemistry: Case Studies of William Cullen and Joseph Priestley." *British Journal for the History of Science*, 21 (68): 1–31.
Gopalakrishnan, P. K. (2000). *Keralathinte Samskarika Charitram*. Trivandrum: State Institute of Languages.
Gough, E. K. (1970). "Cults of the Dead among the Nayars." *Journal of American Folklore*, 71: 446–478.
Graber, O. (1987). *The Formation of Islamic Art*. New Haven, CT: Yale University Press.
Gray, H. and W. H. Lewis (1918). *Anatomy of the Human Body*. Philadelphia, PA: Lea & Febiger.
Great Britain, Parliament, House of Commons, Select Committee on the East India Company, and W. K. Firminger (1969). *The Fifth Report from the*

Select Committee of the House of Commons on the Affairs of the East India Company, 28th July, 1812. New York, NY: A. M. Kelley.

Green, N. (2008). "Moral Competition and the Thrill of the Spectacular." *South Asia Research*, 28 (3): 239–251.

Green, N. (2012). *Making Space: Sufis and Settlers in Early Modern India*. New Delhi: Oxford University Press.

Grube, E. J. and G. Michell (1978). *Architecture of the Islamic World: Its History and Social Meaning*. London: Thames & Hudson.

Guiley, R. (2006). *The Encyclopedia of Magic and Alchemy*. New York, NY: Facts On File.

Gupta, P. (2012). "Monsoon Fever." *Social Dynamics*, 38 (3): 516–527.

Gupta, P., I. Hofmeyr, and M. N. Pearson (2010). *Eyes across the Water: Navigating the Indian Ocean*. Pretoria: Unisa Press.

Habib, I. (1982). *An Atlas of the Mughal Empire: Political and Economic Maps with Detailed Notes, Bibliography, and Index*. Delhi and New York, NY: Centre of Advanced Study in History, Aligarh Muslim University; Oxford University Press.

Hafeel, A., ed. (2007). *Natarivukal: Natuvaidym*. Kottayam: DC Books.

Hamarneh, Sami Khalaf (1986). *History of the Heritage of Medical Sciences of the Arabs and Muslims*. Jordan: Publications of Yarmouk University.

Hamilton, W. (1820). *A Geographical, Statistical, and Historical Description of Hindostan, and the Adjacent Countries: In Two Volumes*. London: John Murray.

Hardiman, D. (2006). *Healing Bodies, Saving Souls: Medical Missions in Asia and Africa*. Amsterdam: Rodopi.

Harrison, M. (1994). *Public Health in British India: Anglo-Indian Preventive Medicine, 1859–1914*. Cambridge, New York, NY: Cambridge University Press.

Harrison, M. (2002). *Climates and Constitutions: Health, Race, Environment and British Imperialism in India*. New Delhi: Oxford University Press.

Hawley, J. S. and D. M. Wulff (1996). *Devī Goddesses of India*. Berkeley: University of California Press.

Hejeebu, S. (2005). "Contract Enforcement in the English East India Company." *The Journal of Economic History*, 65 (2): 496–523.

Heyne, B. (1814). *Tracts, Historical and Statistical, on India*. London: Black Parry.

Hiltebeitel, A. (1988). *The Cult of Draupadī*. Chicago, IL: University of Chicago Press.

Hippocrates and F. Adams (1952). *Hippocratic Writings*. Chicago, IL: Encyclopædia Britannica.

Hippocrates, G. E. R. Lloyd, J. Chadwick, and W. N. Mann (1983). *Hippocratic Writings*. Harmondsworth and New York, NY: Penguin.

Ho, Engseng (2006). *The Graves of Tarim Genealogy and Mobility across the Indian Ocean*. Berkeley: University of California Press.

Hobsbawm, E. J. and T. O. Ranger (1983). *The Invention of Tradition*. Cambridge, New York, NY: Cambridge University Press.

Ḥumaidān, Z. (1995). *A'lām Al-Ḥaḍāra Al-'arabīya Al-Islāmīya Fi 'l-'ulūm Al-Asāsīya Wa't-Taṭbīqīya*, 6 vols. *(Figures in the Arab and Islamic Civilization in the Basic and Applied Sciences)*. Dimašq (Damascus): Manšūrāt Wizārat at-Taqāfa (Ministry of Culture).

Hwa, C. and W. C. Aird (2007). "The History of the Capillary Wall: Doctors, Discoveries, and Debates." *American Journal of Physiology*, 293 (Part 2): H2667–H2679.

Hymavathi, P. (1986). "Health and Hygiene in Medieval Andhra and the Remonstration by Vemana." *Bulletin of the Indian Institute of History of Medicine (Hyderabad)*, 16: 11–18.

Hymavathi, P. (1993). "Religion and Popular Medicine in Medieval Andhra." *Social Scientist*, 21 (1/2): 34–47.

Ibn Abī Uṣaybi'ah and Aḥmad ibn al-Qāsim (1965). *'Uyūn Al-Anbā' fī Ṭabaqāt Al-Aṭibbā' (The Sources and Information of the Classes of Physicians)*, edited by N. Riḍā. Bayrūt (Beirut): Dār Maktabat al-Ḥayāh (Dār al-Ḥayāh library).

Ibn Abī Uṣaybi'ah, Aḥmad ibn al-Qāsim, and 'Uyūn al-Sūd, Muḥammad Bāsil (1998). *'Uyūn Al-Anbā' fī Ṭabaqāt Al-Aṭibbā' (The Sources and Information of the Classes of Physicians)*. Bayrūt: Dār al-Kutub al-'Ilmīyah.

Ibn al-Murtaḍá, A. ibn Y., S. Diwald-Wilzer, and M. Zwettler (1961). *Kitāb Ṭabaqāt Al-Mu'tazilah*. Bayrūt: al-Maṭba'ah al-Kāthūlīkīyah (in kommission bei Frantz Steiner Verlag).

ibn Taghrībirdī, J. a.-D. A. a.-M. Y. (1929–56 [1348–1369 AH]). *Al-Nujūm Al-Zāhirah fī Mulūk Miṣr Wa-Al-Qāhirah*, 12 vols., edited by F. M. Shaltūt. Cairo: Dār al-Kutub al-Miṣrīyah.

Iliffe, J. (1998). *East African Doctors: A History of the Modern Profession*. Cambridge and New York, NY: Cambridge University Press.

Imamuddīn, S. M. (1978). "Māristān (Hospitals) in Medieval Spain." *Islamic Studies*, 17 (1): 45–55.

'Īsā, A. (1930). *Dictionnaire Des Noms Des Plantes En Latin, Français, Anglais Et Arabe = Mu'ğam Asmā' an-Nabāt*. LeCaire: Impr. Nationale.

Jayasuriya, S. d. S. and R. Pankhurst (2003). *The African Diaspora in the Indian Ocean*. Trenton, NJ: Africa World Press.

Jensen, N. T. (2005). "The Medical Skills of the Malabar Doctors in Tranquebar, India, as Recorded by Surgeon T L F Folly, 1798." *Medical History*, 49 (4): 489–515.

Jestice, P. G. (2004). *Holy People of the World: A Cross-Cultural Encyclopedia*. Santa Barbara, CA: ABC-CLIO.

Kahl, O. (2007). *The Dispensatory of Ibn at-Tilmīḏ Arabic Text, English Translation, Study and Glossaries*. Leiden and Boston, MA: Brill.

Kaplan, R. D. (2010). *Monsoon: The Indian Ocean and the Future of American Power*. New York, NY: Random House.

Kauz, R., ed. (2010). *Aspects of the Maritime Silk Road: From the Persian Gulf to the East China Sea*. Wiesbaden: Harrassowitz.

Kelly, K. (2009). *The Middle Ages: 500–1450*. New York, NY: Facts On File.

Khan, M. M. (1997). *Sahih Al-Bukhari*. Riyadh: Darussalam Publications.

Klein, U. and W. Lefèvre (2007). *Materials in Eighteenth-Century Science: A Historical Ontology*. Cambridge, MA: MIT Press.

Kong, Y. C. and D. S. Chen (1996). "Elucidation of Islamic Drugs in Hui Hui Yao Fang: A Linguistic and Pharmaceutical Approach." *Journal of Ethnopharmacology*, 54 (2–3): 85–102.

Kuhn, T. S. (1970). *The Structure of Scientific Revolutions*. Chicago, IL: University of Chicago Press.

Kumar, D. (2006). *Science and the Raj: A Study of British India*. New Delhi, New York, NY: Oxford University Press.

Kurup, K. K. N. (1977). *Aspects of Kerala History and Culture*. Trivandrum: College Book House.

Kutty, P. N. K. M. (2004 [1895]). *Nafeesath Mala*. Calicut: Thirurangadi Book Stall.

Laufer, B. (1925). *Ivory in Chin*. Chicago, IL: Field Museum of Natural History.

Lehmann, A. (1955). *Hallesche Mediziner Und Medizinen Am Anfang Deutsch-Indischer Beziehungen*. Halle: n. p.

Leidy, D. P., W-f. A. Siu, J. Watt, and C. Y. James (1997). *Chinese Decorative Arts*. New York, NY: Metropolitan Museum of Art.

Leslie, C. M. (1976). *Asian Medical Systems: A Comparative Study*. Berkeley: University of California Press.

Leslie, C. M. and A. Young, eds. (1997). *Paths to Asian Medical Knowledge*. Berkeley: University of California Press.

Levey, M. (1973). *Early Arabic Pharmacology. An Introduction Based on Ancient and Medieval Sources*. Leiden: Brill.

Lewis, I. M., A. El Safi, and S. H. A. Hurreiz (1991). *Women's Medicine: The Zar-Bori Cult in Africa and Beyond*. Edinburgh: Edinburgh University Press for the International African Institute.

Li, Shizhen (1990). *Bencao Gangmu*. Beijing: Renmin weisheng.

Lind, J. (1792). *An Essay on Diseases Incidental to Europeans in Hot Climates: With the Method of Preventing Their Fatal Consequences*. London: Printed for J. Murray.

Lipshitz, S. (1978). *Tearing the Veil: Essays on Femininity*. London and Boston, MA: Routledge & Kegan Paul.

Liu, C. and P. Chen (1999). *Well-Known Formulas and Modified Applications*. Beijing: Science Press.

Liyanaratne, J. (1999). *Buddhism and Traditional Medicine in Sri Lanka*. Kelaniya: Kelaniya University Press.

Logan, W. (1951 [1887]). *Malabar Manual*. New Delhi and Madras: Asian Educational Services.

Lynch, L. R. (1969). *The Cross-Cultural Approach to Health Behavior*. Rutherford: Fairleigh Dickinson University Press.

Makhdhum, S. (2012 [1575]). *Fath-Hul-Muin*. Calicut: Poomkavanam Publications.

Makhhdum, S. (1999 [1583]). *Tuhfat-Hul-MujahideenfeeBa-a-Si AkhbarilBurthukhaliyeen*, Translated as *A Tibute to the Warriors with*

Information about Portuguese (Malayalam translated), edited by C. Hamsa. Calicut: Al-Huda Book Stall.
Manucci, N. (1966–67). *Storia do Mogor Or Mogul India*, 4 vols. Calcutta: S. Chand.
Maruyama, M. (1977). *Shinkyū Igaku to Koten no Kenkyū: Maruyama Masao tōyō Igaku rōnshū = (A Study of the Medicine of Acupuncture and Classics: A Collection of Essays on Oriental Medicine by Masao Maruyama)*. Tōkyō: Sōgensha.
Maskiell, M. (2002). "Consuming Kashmir: Shawls and Empires, 1500–2000." *Journal of World History*, 13 (1): 27–65.
Mathers, W. M. and M. Flecker (1997). *Archaeological Report: Archaeological Recovery of the Java Sea Wreck*. Annapolis, MD: Pacific Sea Resources.
Mathew, K. S., T. R. d. Souza, and P. Malekandathil, eds. (2001). *The Portuguese and the Socio-Cultural Changes in India, 1500–1800*. Tellicherry, Kerala: Institute for Research in Social Sciences and Humanities, MESHAR.
Menon, M. (2013). "Making Useful Knowledge: British Naturalists in Colonial India, 1784–1820." Unpublished PhD thesis, University of California, San Diego.
Menon, N. and S. Doshi (1983). *The Performing Arts*. Atlantic Highlands, NJ: Humanities Press.
Menon, T. M. (2000). *A Handbook of Kerala, Volume 2*. Thiruvananthapuram: International School of Dravidian Linguistics.
Meri, J. W. and J. L. Bacharach (2006). *Medieval Islamic Civilization: An Encyclopedia*. New York, NY: Routledge.
Merson, J. (2000). "Bio-Prospecting or Bio-Piracy: Intellectual Property Rights and Biodiversity in a Colonial and Postcolonial Context." *Osiris*, 15 (1): 282–296.
Metcalf, T. R. (2007). *Imperial Connections: India in the Indian Ocean Arena, 1860–1920*. Berkeley: University of California Press.
Meyerhof, M. and P. Johnstone (1984). *Studies in Medieval Arabic Medicine: Theory and Practice*. London: Variorum Reprints.
Miksic, J. N., G. Y. Goh, and S. O'Connor (2011). 2011. *Rethinking Cultural Resource Management in Southeast Asia: Preservation, Development, and Neglect*. London: Anthem Press.
Misra, B. (1969). "Sitala: The Small-Pox Goddess of India." *Asian Folklore Studies*, 28: 133–142.
Miyasita, S. (1976). "A Historical Study of Chinese Drugs for the Treatment of Jaundice." *The American Journal of Chinese Medicine*, 4 (3): 239–243.
Mohanavelu, C. S. (1993). *German Tamilology: German Contributions to Tamil Language, Literature and Culture during the Period 1706–1945*. Madras: Saiva Siddhanta Pathippu Kazhagam.
Mohanty, M. (2004). *Class, Caste and Gender*. New Delhi: Sage.
Morgan, P. (1991). "New Thoughts on Old Hormuz: Chinese Ceramics in the Hormuz Region in the Thirteenth and Fourteenth Centuries." *Iran*, 29: 67–83.
Mottahedeh, N. (2008). *Representing the Unpresentable: Historical Images of National Reform from the Qajars to the Islamic Republic of Iran*. Syracuse, NY: Syracuse University Press.

Muhammad, Q. (2000 [1607]). *Muhiyuddeen Mala*. Calicut: Thirurangadi Book Stall.
Mukerji, C. (2009). *Impossible Engineering: Technology and Territoriality on the Canal Du Midi*. Princeton, NJ: Princeton University Press.
Mukerji, C. (2010). "The Territorial State as a Figured World of Power: Strategics, Logistics, and Impersonal Rule." *Sociological Theory*, 28 (4): 402–424.
Mukharji, P. (2011). "Lokman, Chholeman and Manik Pir: Multiple Frames of Institutionalising Islamic Medicine in Modern Bengal." *Social History of Medicine*, 24 (3): 720–738.
Musliar, A. B. (2001 [1885]). *Fee Shifau-n-Nasi: Ithu Orumichu Koottappetta Pazhaya Upakaram Tarjama Kitab ("This Translated Compilation Contains Remedies for People")*. Thirurangadi: C. H. Muhammed Koya.
Muzur, A., A. Skrobonja, V. Rotschild, and A. Skrobonja (2005). "Saints-Protectors from Snake Bite: A Short Overview and a Tentative Analysis." *Journal of Religion and Health*, 44 (1): 31–38.
Namboodiripad, E. M. S. (2010). *History, Society, and Land Relations: Selected Essays*. New Delhi: LeftWord Books.
Nambootiri, M. V. V. (1979). *Mantrikavidyayum Mantravadappattukalum*. Kottayam: DC Books.
Nambothiri, S. (1990). *Chikitsa Manjari* (translated from-Malayalam). Alapuzha: Vidyarambham.
Naraindas, H. and C. Bastos (2011). "Healing Holidays? Itinerant Patients, Therapeutic Locales and the Quest for Health." *Anthropology and Medicine*, 18 (1): 1–6.
Naraindas, H. and C. Bastos (2015). *Healing Holidays: Itinerant Patients, Therapeutic Locales and the Quest for Health*. London: Routledge.
Narayanan, M. G. S. (1972). *Cultural Symbiosis in Kerala*. Trivandrum: Kerala Historical Society.
Nasr, S. H. (1987). *Islamic Art and Spirituality*. Albany: State University of New York Press.
Needham, J. (1969). *The Grand Titration: Science and Society in East and West*. London: Allen & Unwin.
Needham, J. and Ling Wang (1956). *Science and Civilisation in China, Volume 2*. Cambridge: Cambridge University Press.
Nicholas, R. W. (1981). "The Goddess Sitala and Epidemic Smallpox in Bengal." *The Journal of Asian Studies*, 41 (1): 21–45.
Ogborn, M. (2007). *Indian Ink Script and Print in the Making of the English East India Company*. Chicago, IL: University of Chicago Press.
Osella, F. and C. Osella (2003). "Migration and the Commoditisation of Ritual: Sacrifice, Spectacle and Contestations in Kerala, India." *Contributions to Indian Sociology*, 37 (1/2): 109–139.
Paige, K. and J. M. Paige (1981). *The Politics of Reproductive Ritual*. Berkeley: University of California Press.
Pallath, J. J. (1995). *Theyyam: An Analytical Study of the Folk Culture, Wisdom, and Personality*. New Delhi: Indian Social Institute.

Pearce, J. M. S. (2007). "Malpighi and the Discovery of Capillaries." *European Neurology*, 58 (4): 253–255.
Pearson, M. N. (1995). "The Thin End of the Wedge. Medical Relativities as a Paradigm of Early Modern Indian-European Relations." *Modern Asian Studies*, 29 (1): 141–170.
Pearson, M. N. (2001). "Hindu Medical Practice in Sixteenth-Century Western India: Evidence from Portuguese Sources." *Portuguese Studies*, 17: 100–113.
Pearson, M. N. (2005). *The World of the Indian Ocean, 1500–1800: Studies in Economic, Social, and Cultural History*. Burlington, VT: Ashgate.
Pearson, M. N. (2006). "Portuguese and Indian Medical Systems: Commonality and Superiority in the Early Modern Period." *Revista De Cultura*, 20: 116–141.
Pearson, M. N. (2011). "Medical Connections and Exchanges in the Early Modern World." *PORTAL: Journal of Multidisciplinary International Studies*, 8 (2): 1–15. http://epress.lib.uts.edu.au/journals/index.php/portal/article/view/1643/2548.
Pfleiderer, B. (2006). *The Red Thread: Healing Possession at a Muslim Shrine in North India*. Delhi: Aakar Books.
Phillips, T., trans. (1717). *An Account of the Religion, Manners, and Learning of the People of Malabar in the East-Indies in Several Letters…* London: Black Swan.
Porkert, M. (1973). *The Theoretical Foundations of Chinese Medicine: Systems of Correspondence*. Cambridge: MIT Press.
Pormann, P. E. and E. Savage-Smith (2007). *Medieval Islamic Medicine*. Washington, DC: Georgetown University Press.
Prange, S. R. (2008). "Scholars and the Sea: A Historiography of the Indian Ocean." *History Compass*, 6 (5): 1382–1393.
Radwanski, S. A. and G. E. Wickens (1981). "Vegetative Fallows and Potential Value of the Neem Tree (*Azadirachta indica*) in the Tropics." *Economic Botany*, 35 (4): 398–414.
Raj, K. (2007). *Relocating Modern Science: Circulation and the Construction of Knowledge in South Asia and Europe, 1650–1900*. Houndmills, Basingstoke, Hampshire; and New York, NY: Palgrave Macmillan.
Rajan, V. (1980). *Agathiyar Irrandayiram*. Thanjavur: Don Bosco Press.
Ramakrishnan, P. S., K. G. Saxena, and U. M. Chandrashekara, eds. (1998). *Conserving the Sacred: For Biodiversity Management*. Enfield, NH: Science.
Raman, B. (2008). "The Familial World of the Company's Kacceri in Early Colonial Madras." *Journal of Colonialism and Colonial History*, 9 (2). doi:10.1353/cch.0.0011.
Raman, B. (2012). *Document Raj: Writing and Scribes in Early Colonial South India*. Chicago, IL: University of Chicago Press.
Rangan, H., T. Denham, and J. Carney (2012). "Environmental History of Botanical Exchanges in the Indian Ocean World." *Environment and History*, 18 (3): 311–342.
Raṇṭattāṇi, H. (2007). *Mappila Muslims: A Study on Society and Anti Colonial Struggles*. Calicut: Other Books.

Robinet, I. and P. A. Wissing (1990). "The Place and Meaning of the Notion of Taiji in Taoist Sources Prior to the Ming Dynasty." *History of Religions*, 29 (4): 373–411.
Robinson, S. (1969). *A History of Printed Textiles*. Cambridge: MIT Press.
Rotondo-McCord, L. E. and R. D. Mowry (2000). *Heaven and Earth Seen Within: Song Ceramics from the Robert Barron Collection*. New Orleans: New Orleans Museum of Art.
Roxburgh, W. and A. Anderson (1810). "Papers in Colonies and Trade." *Transactions of the Society, Instituted at London, for the Encouragement of Arts, Manufactures, and Commerce*, 28: 249–316.
Russell, P. (1796). *An Account of Indian Serpents, Collected on the Coast of Coromandel; Containing Descriptions and Drawings of Each Species; Together with Experiments and Remarks on Their Several Poisons*. London: Printed by W. Bulmer and Co. for George Nicol.
Sadasivan, S. N. (2000). *A Social History of India*. New Delhi: Ashish Publishing House.
Sahl, i. S. and O. Kahl (2003). *The Small Dispensatory*. Leiden and Boston, MA: Brill.
Schaffer, S. (2003). "Enlightenment Brought Down to Earth." *History of Science*, 41: 257–268.
Schottenhammer, A. (2005). *Trade and Transfer across the East Asian "Mediterranean"*. Wiesbaden: Harrassowitz.
Schumaker, L., D. Jeater, and T. Luedke (2007). "Histories of Healing: Past and Present Medical Practices in Africa and the Diaspora." *Journal of Southern African Studies*, 33: 707–714.
Scott, R. and J. Guy, eds. (1995). *South East Asia and China: Art, Interaction and Commerce: Colloquies on Art and Archaeology in Asia, no. 17*. London: University of London, Percival David Foundation of Chinese Art.
Selin, H. (1997). *Encyclopaedia of the History of Science, Technology, and Medicine in Non-Western Cultures*. Dordrecht and Boston, MA: Kluwer Academic.
Sengers, G. (2003). *Women and Demons: Cult Healing in Islamic Egypt*. Leiden and Boston, MA: Brill.
Şeşen, R., C. Akpınar, E. İhsanoğlu, and C. İzgi (1984). *Catalogue of Islamic Medical Manuscripts in the Libraries of Turkey*. İstanbul: Research Centre for Islamic History, Art and Culture.
Sezgin, F. (1970). *Geschichte Des Arabischen Schrifttums 3: Medizin, Pharmazie, Zoologie, Tierheilkunde—Bis Ca. 430 H*. Leiden: Brill.
Sezgin, F. (1996). *Galen in the Arabic Tradition, Texts and Studies*, Vol. 1. Frankfurt am Main: Institute for the History of Arab-Islamic Science.
Sezgin, F. and M. Amawi (1996). *Ḥunayn Ibn Isḥāq = Ḥunain b. Isḥāq (d. 260/873): Texts and Studies*. Frankfurt am main: Institute for the History of Arabic-Islamic Science.
Shapin, S. (1974). "The Audience for Science in Eighteenth Century Edinburgh." *History of Science*, 12: 95–121.
Shulman, D. D. and V. N. Rao (1995). *Syllables of Sky: Studies in South Indian Civilization*. Delhi and New York, NY: Oxford University Press.

Simpson, E. and K. Kresse (2008). *Struggling with History: Islam and Cosmopolitanism in the Western Indian Ocean.* New York, NY: Columbia University Press.
Singh, K. S. (2002). *People of India: Introduction.* New Delhi, Oxford and New York, NY: Oxford University Press.
Singh, S. B. (1966). *European Agency Houses in Bengal, 1783–1833.* Calcutta: Firma K. L. Mukhopadhyay.
Sivin, N. (1988). "Science and Medicine in Imperial China—The State of the Field." *The Journal of Asian Studies,* 47 (1): 41–90.
Smith, F. M. (2006). *The Self Possessed Deity and Spirit Possession in South Asian Literature and Civilization.* New York, NY: Columbia University Press.
Snell, R. (1992). *Clinical Anatomy, Upper and Lower Extremities,* translated into Arabic by A. Khatib and R. Amid. Damascus, Syria: Dar al-Shady.
Spary, E. C. (2000). *Utopia's Garden: French Natural History from Old Regime to Revolution.* Chicago, IL: University of Chicago Press.
Speziale, F., ed. (2012). *Hospitals in Iran and India, 1500–1950s.* Leiden and Boston, MA: Brill.
Sreedhara Menon, A. (2007). *A Survey of Kerala History.* Kerala, India: DC Books.
Sreedhara Menon, A. (2008). *Cultural Heritage of Kerala.* Kottayam: DC Books.
Sreedhara Menon, A. and K. K. Kusuman (1990). *A Panorama of Indian Culture: Professor A. Sreedhara Menon Felicitation Volume.* New Delhi: Mittal.
Stephens, C., J. Porter, C. Nettleton, and R. Willis (2006). "Disappearing, Displaced, and Undervalued: A Call to Action for Indigenous Health Worldwide." *Lancet,* 367 (9527): 2019–2028.
Stern, P. J. (2011). *The Company-State: Corporate Sovereignty and the Early Modern Foundations of the British Empire in India.* New York, NY: Oxford University Press.
Strathern, A. and P. J. Stewart (2005). *Contesting Rituals: Islam and Practices of Identity-Making.* Durham, NC: Carolina Academic Press.
Subrahmanyam, S. (2005). *From the Tagus to the Ganges.* New Delhi: Oxford University Press.
Subramanian, P. (1989). *Tañcai marāṭṭiya maṉṉar mōṭi āvaṇat Tamiḻakamuṉ Kurippuraiyum Mutal Tokuti (Thanjāvur Maratha Kings' Modi Documents of Tamilakam and Notes), Volume I.* Thanjavur: Tamil University.
Sudhoff, K. (1926). *Essays in the History of Medicine.* New York, NY: Medical Life Press.
Taylor, G. (2010). "The Kitab Al-Asrar: An Alchemy Manual in Tenth-Century Persia." *Arab Studies Quarterly,* 32 (1): 6.
Thasarathan, A., V. S. Subbaraman, S. Hikosaka, and G. John Samuel (1993). *A Descriptive Catalogue of Palm-Leaf Manuscripts in Tamil.* Madras: Institute of Asian Studies.
Thurston, E. and K. Rangachari (1909). *Castes and Tribes of Southern India,* 7 vols. Madras: Government Press.

Toussaint, A. (1961). *Histoire De l'Océan Indien*. Paris: Presses universitaires de France.
Travers, R. (2007). *Ideology and Empire in Eighteenth Century India: The British in Bengal, 1757–93*. New York, NY: Cambridge University Press.
Tripathi, A. (1956). *Trade and Finance in the Bengal Presidency, 1793–1833*. Bombay: Orient Longmans.
Trostle, J. A. (2005). *Epidemiology and Culture*. Cambridge, UK; and New York, NY: Cambridge University Press.
Turner, B. S. and Y. Zheng (2009). *The Body in Asia*. New York, NY: Berghahn Books.
Unschuld, P. U. (2010). *Medicine in China: A History of Ideas*. Berkeley: University of California Press.
Uragoda, C. G. (1987). *A History of Medicine in Sri Lanka from the Earliest Times to 1948*. Colombo: Sri Lanka Medical Association.
Vaidyar, M. K. (1951). *Mahasaram*. Madras: Government Oriental Manuscripts Library.
Van Toller, S. and G. H. Dodd, eds. (1992). *Fragrance: The Psychology and Biology of Perfume*. New York, NY: Elsevier Applied Science.
Varier, N. V. K. (2002). *Ayurveda Charithram*. Kottakkal: Arya Vaidya Sala.
Varthema, L. d., J. W. Jones, and G. P. Badger (1863). *The Travels of Ludovico Di Varthema in Egypt, Syria, Arabia Deserta and Arabia Felix, in Persia, India, and Ethiopia, A.D. 1503 to 1508*. London: Printed for the Hakluyt Society.
Verelst, H. (1772). *A View of the Rise, Progress, and Present State of the English Government in Bengal Including a Reply to the Misrepresentations of Mr. Bolts, and Other Writers*. London: J. Nourse.
Vink, M. P. M. (2007). "Indian Ocean Studies and the 'New Thalassology.'" *Journal of Global History*, 2 (1): 41–62.
Visscher, J. C. and H. Drury (1862). *Letters from Malabar*. Madras: Printed by Gantz, at the Adelphi Press.
Wang, Z., C. Ping, and X. Xing (1999). *History and Development of Traditional Chinese Medicine*. Beijing, Amsterdam, and Tokyo: Science Press, IOS Press, Ohmsha.
Wangu, M. B. (2003). *Images of Indian Goddesses: Myths, Meanings, and Models*. New Delhi: Abhinav.
Webster, T. (2005). "An Early Global Business in a Colonial Context: The Strategies, Management, and Failure of John Palmer and Company of Calcutta, 1780–1830." *Enterprise and Society*, 6 (1): 98–133.
Williams, D. R. (2005). *The Other Side of Zen: A Social History of Sōtō Zen: Buddhism in Tokugawa Japan*. Princeton, NJ: Princeton University Press.
Wirgin, J. C. (1970). *Sung Ceramic Designs…: Jan C. Wirgin*. Göteborg: Elanders.
Wiseman, N. and A. Ellis (1996). *Fundamentals of Chinese Medicine = Zhōng Yī Xué Jī Chǔ*. Brookline, MA: Paradigm.
Wiseman, N. and Ye Feng (1998). *A Practical Dictionary of Chinese Medicine*. Brookline, MA: Paradigm.

Wujastyk, D. and F. M. Smith (2008). *Modern and Global Ayurveda Pluralism and Paradigms*. Albany: State University of New York Press.
Yalman, N. (1963). "On the Purity of Women in the Castes of Ceylon and Malabar." *The Journal of the Royal Anthropological Institute of Great Britain and Ireland*, 93 (1): 25–58.
Yāqūt i. 'A. A. a.-H. and D. S. Margoliouth (1923). *The Irshád Al-Aríb Ilá ma'rifat Al-Adíb: Or, Dictionary of Learned Men of Yáqút*. London: Luzac.
Zhao, R. (2012 [1911] [circa 1200]). *Chau Ju-Kua: His Work on the Chinese and Arab Trade in the Twelfth and Thirteenth Centuries Entitled Chu-Fan-Chï*, translated by W. W. Rockhill and F. Hirth. Hong Kong: Forgotten Books.
Ziegenbalg, B. and W. Caland (1926). *Ziegenbalg's Malabarisches Heidenthum*. Amsterdam: Koninklijke Akademie van Wetenschappen.
Ziegenbalg, B. and W. Germann (1869). *Genealogy of the South-Indian Gods: A Manual of the Mythology and Religion of the People of Southern India, Including a Description of Popular Hinduism*. Madras: Higgenbotham.
Županov, I. G. (2005). *Missionary Tropics: The Catholic Frontier in India, 16th–17th Centuries*. Ann Arbor: University of Michigan Press.
Zysk, K. G. (1991). *Asceticism and Healing in Ancient India: Medicine in the Buddhist Monastery*. New York, NY: Oxford University Press.

Contributors

Mahmud Angrini is Physician specializing in Laboratory Medicine and a Former Health Manager of the International Medical Corps, Turkey Office. He holds masters degrees in Laboratory Medicine, History of Medicine, and History of Sciences and Techniques. He is currently studying for a masters degree in Molecular and Cellular Biology, University Pierre et Marie Curie, Paris.

P. K. Yasser Arafath is Assistant Professor in History at Delhi University. His areas of specialization include History of Science and Technology, South Indian History, Cultural History of Islamic Communities, and Vernacular History. He has published peer reviewed journal articles and written for newspapers and magazines.

Minakshi Menon is Postdoctoral Fellow at the Max Planck Institute for the History of Science, Berlin. She is currently working on a book manuscript titled "Empiricism's Empire: Natural-Knowledge Making, State Making and Governance in East India Company India, 1784–1857." She received her PhD in History (Science Studies) from the University of California San Diego. Her interests include the history of colonial science, early-modern natural history, South Asian history, and the history of British imperialism.

Lisa C. Niziolek is postdoctoral researcher in East and Southeast Asian Archaeology at The Field Museum of Natural History in Chicago, where she investigates early maritime trade networks in the South China Sea region. Her doctoral research focused on the organization of ceramic production in the prehispanic Philippines. She has conducted fieldwork in the Philippines, Puerto Rico, Ireland, Spain, and the United States and has extensive experience in museum anthropology and collections-based research.

Amanda Respess is National Science Foundation Graduate Research Fellow and doctoral pre-candidate at The University of Michigan's Joint Program in Anthropology & History. Previously, as a Boone Scholar Intern at The Field Museum of Natural History, she conducted research on the 13th-century *Java Sea Shipwreck* collection. Her current research focuses on the medieval exchange of medical knowledge between China and Iran.

S. Jeyaseela Stephen is Directeur, Institut pour études Indo-Européennes. He was Professor of Maritime History (2001–2013) at Visva-Bharati University, Santiniketan, West Bengal, India. He is the author of numerous books on maritime history of early modern India including Oceanscapes: *Tamil Textiles in the Early Modern World* (2014) and *European and Tamil Encounters in Modern Sciences, 1507–1857* (2015). He received the best book prize of the year 1999 from the Government of Tamil Nadu.

Facil Tesfaye is Assistant Professor and Director of the African Studies Programme at the University of Hong Kong. Although his areas of specialization is African History, he is currently working on the History of Medicine in certain parts of Africa including Indian ocean Africa. He has published peer reviewed journal articles and his recent book "Statistique(s) et Genocide au Rwanda" came out in 2014.

Anna Winterbottom is British Academy Postdoctoral Fellow at the University of Sussex, UK. She is author of *Hybrid Knowledge in the Early East India Company World*, co-editor of *The East India Company and the Natural World*, and has published several journal articles and book chapters on topics related to the history of science and medicine, the Indian Ocean region, and colonial history.

Index

"A New Sure, Short, and Easy
 Method of Treating Persons
 Affected by Rabies," 128
 see also du Choiseul, Jean-Baptiste
abortion, 111
Abu'l-Fazl, 'Allami (1551–1602),
 111
*An Account of Indian Serpents
 Collected on the Coast of
 Coromandel*, 139
 see also Russell, Patrick
Acorus calamus, 131
Adams, Julia, 172
*al-Adkar Min kalam Sayyad
 al-Abraar*, 107
 see also Imam Nawawi
Agasthiyar vaythyama Ayinooru, 136
Agastiyar Irandaayiram, 15, 130–2
 see also Grundler
Ainslie, Whitelaw (1767–1837), 16,
 136
alangi root, 133
Alavi, Seema, 6, 9
Alberuni. *See* al-Biruni
Ali, Muhammad (1769–1849), 18
Alpers, Edward, 5
ambergris of Azania, 2
*Anmækning Om mallebar Loegernes
 Kundskab I Chirurgien*, 132
 see also Folly, Theodor Ludvig
 Frederik
al-Aqrabadhin, 49
 see also Sabur Ibn Sahl

Arabic, 12, 90, 100
 Arabic-Malayalam texts, 100, 108
 cosmopolis, 109, 120
 letters, 14, 108
 medical texts, 42, 45, 72, 107
 numerals, 91, 108, 110
 pharmacology, 6, 49
 plant names, 72, 136
 translations of Galen, 42, 54
 works of al-Razi, 38, 56
Armenian bole, 2
Arnold, David, 125
arsenic, 15, 138–9, 143
Artemisia afra, 25
arthralgia, 56
Aśoka, Emperor (c. 259–222 BC), 6
Attwell, Guy, 6, 9
Avicenna. *See* Ibn Sina
ayn (evil eye), 106–7
Ayurveda, 9, 26, 99, 126, 130, 14
 see also medicine
Azadirachta indica, 3

Baghdad, 6, 37, 109
 hospital (*Al-Bimaristan
 al-Aodedi*), 37
Baker, George, (fl. 1792), 171
Balfour, Edward (1813–1889), 112
al-Balkhi, Abu al-Hasan Shahid ibn
 al-Husayn (d. 936), 38
al-Balkhi, Abu al-Qasim al-Ka'bi
 (d. 931), 37
al-Balkhi, Abu Zayd (c. 849–934), 37

Banks, Sir Joseph (1743–1820), 157, 162
Barbosa, Duarte, 101–2, 110, 112, 116
Bastos, Cristiana, 11, 19, 23
Basu, Helen, 5
bathing, 7, 52, 116
  see also *Hammam*
Bencao Gangmu, 73
Bengal, 102, 152, 159, 164
Bhagavatis, 100
Bhattacharyya, Anoushka, 11, 19
al-Biruni, Abi Rayhan (973–1048), 5, 38–9
Boag. W. (fl. 1809), 139
borax, 2
bori (medicine cult), 5
Bose, Sugata, 20
boteh (paisley), 85–6
Bowring, John (1792–1872), 18
Braudel, Ferdinand, 2, 9
Brazil, 26
British Empire, 8, 11, 125
Buddhism, 3, 13–14, 75, 85

Cairo, 6, 7, 18
Calcutta, 21, 22, 138, 157, 162
Caldwell, Sarah, 102
Calotropis gigantean, 132
Campbell, Sir Archibald (1767–1843), 137, 157, 166
Cannabis sativa, 135
Canon on Medicine. See *Qanun*
celandine, 2
Ceylon. *See* Sri Lanka
Chakrabarti, Pratik, 17, 125, 136
chank lime, 135
Chaudhuri, K. N., 2
Cheerma Bhagavati, 101
Cheranellur Durga, 101–2
chicken pox (*Ponganpani*), 101
chikungunya, 1, 24
childbirth, 13, 73, 100, 103, 106, 110, 115, 116, 118
  see also fertility
China, 1, 3, 6, 9, 12–13, 63–93, 159

Chipman, Leigh, 6
cholera, 1, 22
Chottanikkara temple, 101
Chowdhury, Rashed, 11, 18
Christianity, 13, 15, 18, 103, 105, 127
cinnamon, 2, 162
Circars, 170
Clot, A. B. (1793–1868), 18
cloves, 162
Cochran, Sherman, 9
Coeurdoux, Gaston Laurent (1691–1779), 127
coffee, 162
Cook, Harold J., 8, 16, 125, 140, 154
Copenhagen, 132
Corconda, 162
Cornwallis, Lord Charles (1738–1805), 158, 164, 168
Coromandel Coast, 15, 125, 127, 139
credit, 16
Critchlow, 88
Cullen, William (1710–1790), 153
Curcuma domestica, 133

da Gama, Arthur Ignacio, 22–3
da Silva, Ezequiel, 22–3
Dale, Stephen, 105
Dalrympyle, Alexander (1737–1808), 17, 156, 160, 169
de Bils, Louis (1624–1669), 154
dengue, 1
Dioscorides, 6
disease, 1, 20–4, 53–4, 126, 129–35, 142
  tropical, 144
Duffin, William, (fl.1788), 137
Duncan's Medical Commentaries, 138
Dutch East India Company. See *Vereenigde Oost-indische Compagnie* (VOC)
Dutch Physicians, 125

Eberhard, 83
Edinburgh, 152–62, 167
  medical school, 152–3

Index

Egypt, 18
Egyptian opium, 2
El Said, Issam, 88
Elgood, Cyril, 6
English East India Company (EIC), 15, 17, 126, 134, 137–9, 163, 170
Company-State, 151, 172, 173 (*see also* Stern, Philip)
epidemics, 18, 21, 22, 102
Ethiopia, 11, 19
Europe, 9, 125
European society, 141
Europeans, 142
evil eye. *See* ayn

al-Fakher, 38
see also al-Razi
Fardos al-Hekma, 46
see also al-Tabari
Fath-hul-Muin, 100, 109, 114, 116, 117
Fawaid, 107. *See also* Sharji', Imam
Fee Shifau-n-Nasi, 100, 106
Feodorovna, Empress Maria (1847–1928), 19
fertility, 13, 64, 84, 100, 106–12
fertility healing, 119
fertility rituals, 14, 102, 104, 116
fertility symbolism, 83–4
goddess, 101–2, 118
infertility, 101, 102, 109
(*see also* medicine)
maleyankettu, 117
masculine fertility/reproductive psyche, 114
medicine, 87
midwifery, 117
sexual energy, 112
sexual hygiene, 114–16
see also snakes
fever, 48, 131, 135
hospital, Calcutta, 22
treatment, 68, 138
fever bark (*Swietenia febrifuga*), 162

fi Awjaa'al Mafasel, 11, 38–58
see also al-Razi, Abu Bakr Muhammed ibn Zakariya
firangi, 1
see also syphilis
Folly, Theodor Ludvig Frederich, 132–3
Fort St. George at Madras, 138, 144
France, 128
Freeman, Rich, 101
fuke, 64, 66, 90
fuke formulas, 78
see also women's medicine
Furber, Holden, 171

Galen, 42, 52–4
Galenic medicine, 45, 73, 83, 88–9
German Doctors, 132
al-Ghazali, Abu Hamid Muhammad ibn Muhammad (1058–1111), 114
ginger, 2
Goa, 11, 22, 23
Godavari river, 167, 169–70
Green, Nile, 4
Grundler, Johann Ernst (1677–1720), 15, 16, 130–1

Hadrami society, 4
Haliburton, David, 164
Halle, 130–1
Hammam, 7, 53
al-Hawi, 53–7
see also al-Razi
Hayne, Benjamin (fl. 1790), 16, 128, 134
healing, 5, 10, 23–7, 99, 104, 119
amulets, 79
evil spirits, 103
fertility prayers, 118
rituals, 103
Sihr healers, 110
Zulu and Xhosa healing practices, 25
see also fertility healing, medicine

Hinduism, 3, 13, 135
  healing, 5
  moorkhan, 112
  rituals, 13
  spirit possession, 5
Hippocrates, 52
hookworm infection, 20, 24
Hui Hui Yaofang, 72–3, 90
Humaidan, Zuhair, 38
Hurgobin, Yoshina, 11, 18, 19, 21

Ibn Abi Usaybi'ah (1203–1270), 39, 42
Ibn al-Tilmidh, 49
Ibn Hajar (1372–1449), 114
Ibn Qayyim Al-Jawziyya (1292–1350 AD), 107
Ibn Sina (980–1037), 7, 12, 37, 49, 73
  see also Avicenna
Imam Nawawi (1233–1277), 107, 114
Imam Reza (765–817), 4
Imam Sharji' (1410–1488), 107
Imam Subuki, 114
India, 4, 7, 8, 19, 111, 126, 144, 159, 163
Indian Ocean World
  colonial presence, 8, 106
  commerce, 153, 155
  cultural exchange, 91
  medicine (*see* medicine, *materia medica*)
  migration, 4, 9, 27
  theoretical formulation, 2
  trade, 2, 12, 67–8, 71–2, 133, 152, 155, 157, 158–9, 164, 165
  trade networks, 91, 125
indigo, 17, 155, 158–62, 173
  see also nerium
International Center on the Indigenous Phytotherapy Studies, 26
Iran, 86
  see also Persia
iron oxide, 2
Islam, 3, 109

Islamic humoral theory, 12, 44, 83
Islamic medicine, 4, 72, 77, 88
  see also medicine, *Tibb-un-Nabi* (prophetic medicine)
ivory, 68
ivory horn, 2
ivory powder, 13, 69, 70–1

Jahangir, Emperor (1569–1627), 8
James Anderson (1738–1809), 138, 171
James Lind (1716–1794), 144
Jansen, Karine, 11, 22, 24
Java, 12
Java Wreck, 63, 66–77
Jean-Baptiste du Choiseul, Friar, 127
Jin Gui Yao Lue Fang Lun, 66, 70
  see also Zhang Zhongjing
Jones, William, 138

Kajiwara Shozen, 3
Kali, 101
Kalpasthanam, 134–5
Kannur, 104
Kavadi, Shirish, 11, 20
al-Kayyal, Ahmad ibn, 38
Kerala, 99, 102, 109
  Calicut, 105, 112
  Kasaragode, 104
  medieval Kerala, 116
  Panthalayani, 112
Kerala Panchakarma therapy, 1
King Mahinda IV (956–972 AD), 6
King Sena II (851–885 AD), 6
Kitab al-Mansuri fi al-tibb, 38, 55
  see also al-Razi
Kitab al-Mayamiri, 54
  see also al-Razi, Galen
Kitab al-saidla fi al-tibb, 6
  see also al-Biruni
Kitab Ikhbar al-'ulama' bi-akhbar al-hukama, 39
  see also al-Qifti
Kodungallor temple, 101

Index 201

Komaramas, 101
Korakkar Siddhar, 134
Krishna river, 167, 169–70
Kyd, Colonel Robert (1746–1793), 171

La Rue, Michael, 18
Laplante, Julie, 11, 25
lemon plant, 133
leprosy, 111
Lieutenant Lennon, 170
lingua franca, 5
London, 135–6, 139, 157–8, 164
long gu, 68
lotus plant, 85

Maa alfark, 56
  see also al-Razi
Madras, 7, 15, 134, 136, 137, 152, 157
Madras Courier, 138
Mahishasura Mardhini, 101
Makhdhum, 111
Malabar, 103
Malabar Coast, 13, 99, 109
malachite, 2
malaria, 1, 22
  see also fever
Malek Library, 40–1
Malpighi, Marcello (1628–1694), 45
al-Mansuri, 43
  see also al-Razi
Maqala fi al-naqras, 38–9
  see also "On Gout" and al-Razi
Maritime Silk Road, 12, 63
Mariyamma, 101
  see also healing goddess
al-Masam'i, Ahmad ibn al-Hasan, 37
Mascarenes, 11
  Mauritius, 11
  Reunion, 24
Materia medica, 2, 3, 13, 14, 15, 23, 68, 72, 84, 136, 143
"Materia Medica of Hindoostan," 136
Mazeret, Alexis, 127

medicine
  allopathic, 7, 9, 26
  alternative, 9
  biomedicine, 10, 25
  chemical revolution, 9
  Chinese Tibetan medicine (*Sowa rigpa*), 9
  colonial, 8, 9, 100
  Galenic (*see* Galen, Galenic medicine)
  Greco-Islamic medicine, 109
  Greco-Roman medicine, 109
  humoral theory, 9, 11, 44–8
    Hippocratic theory of humors, 12
  Indian (*see* ayurveda, siddha, *unani tibb*)
  Indo-Islamic medicine, 109
  magico-medicine, 110
  Mantra, 116
  medical tourism, 10
  missionary, 8
  non-Western, 8
  prophetic medicine (see *Tibb-un-Nabi*)
  Qu'anic medicine, 110
  sexual imbalance, 13, 70
  Siddha medicine, 127
    founding saints, 133
  South African traditional medicine (*see* muti)
  Thanjavar pills, 15
  theoretical formulations of al-Razi, 37–58
  Western, 8, 9, 23
  women's medicine, 13, 69, 82, 92
  Ying-yang theory, 12, 66–7, 73, 79, 82
melon stalk powder, 83
Menelik, Emperor (r. 1889–1913), 19
Menon, Minakshi, 11, 14–17
mercury from China, 133
mercury sublimation, 132–3
Middle East, 78
Mihintale hospital, 6

missionaries
  Jesuit, 15, 127–8, 130
  Protestant, 15, 136
  Tranquebar, 15, 101, 128, 130, 132
monsoon, 1, 27, 167
Moringa pterygosperma, 132
Mozambique, 11
Mughal Empire, 7–8
Muhammed, Qazi (1893–1947), 105
Muhiyuddheenmala, 100, 119
  see also Arabic-Malayalam texts
Mukerji, Chandra, 155
muti, 9, 10, 25
myrrh, 73

Nafesathmala, 100, 105–6
al-Nafīs, Abū'l Hasan 'Alī ibn (al-Qurashī) (d. 1288), 7
Nerium, 17, 160, 169, 172
  see also fever bark
Nestorian Christians, 6, 12, 72, 90
Nishapur, 86
Nobili, Roberto (1577–1656), 127
nutmeg, 2, 162

Oakley, Charles, 164, 168, 171
"On Joint Pains." See *fi Awjaa' al-Mafasel*. See also al-Razi
Oriental Repertory, 156, 169
  see also Alexander Dalrymple
Oweis, 87

palaiyakkarar, 17, 161
Parham, Cyrus, 85
Paris, 128
Parman, 88
Pasteur, Louis (1822–1865), 15, 128
Patna asylum, 19
Payyanur *Kavu*, 101, 103
Pearl Fishery Coast, 125
Pearson, Michael, 8, 10, 27
peony, 85
Persia, 73
Persian, 6, 12, 90
  cosmopolis, 109, 120

decorative motifs, 79, 84–6
medical texts, 6, 9, 38
medicine, 2, 12, 71, 72
plant names, 6, 136
traders, 72
Phyllanthus emblica, 132
plague, 1, 18, 22
  see also Egypt, epidemics
poison, 16, 112–13, 127, 128, 131, 137, 139
  see also snakes, toxicology
pomegranates, 79, 84
Ponanni Nalakath Kunji Moideen Kutty, 106
Pondicherry, 127–8
Portuguese imperial movement, 11
prasadam, 14
Prince Serfoji (1777–1832), 158
Prognostics, 54
Puthenkavu Bhagavati, 102

Qala'un, Sultan of Egypt, 7
Qanun, 7, 12, 72–3
  see also Ibn Sina
al-Qifti, Jamaluddin (1172–48), 42
Quanzhou, 63, 67, 72, 90
Qu'ran, 107, 110

rabies, 15, 128
Raj, Kapil, 8, 16
al-Razi, Abu Bakr Muhammed ibn Zakariya (854–925), 7, 11, 37–9
  see also Arabic works of al-Razi, *al-Fakher, fi Awjaa' al-Mafasel, al-Hawi, Kitab al-Mansuri fi al-tibb, Kitab al-Mayamiri*
Read, Captain Alexander, 166
rhubarb, 2
Rifayee mala, 106
  see also Sheikh Rifayee
Risalat Abi Rayhan fi fihrist kutub al-Razi, 39
Risaleh fi al-Hasba and al-Jadari, 38

Roque, Ana, 11, 19, 23
Ross, Andrew (d.1797), 17, 152–74
Rottler, Johann Peter (1749–1836), 136
Roxburgh, William (1751–1815), 17, 152–74
Royal Society of London, 138
Rufus of Ephesus, (98–117), 42
Russell, Patrick (1726–1805), 138–9, 164, 167
Russian Empire, 18
Russian Red Cross, 19
Ryukyu islands, 3

Sabur Ibn Sahl (d. 869), 49
al-Sadiq, Ja'far, 84
Safavid empire, 7
Samalkota, 166
Schwartz, Christian Fredrick (1726–1798), 136
sciatica, 46, 47
sesamum orientale, 131
Sesbania grandiflora, 132
sexual symbolism, 83–4
  see also fertility symbolism
Shang Han Lun, 66–7
  see also Zhang Zhongjing
Sheikh Muhiyuddheen, 108
Sheikh Rifayee (118–1182), 106
Sheikh Sainduddin Makhdum, 114, 119
al-Shokouk Ala Jalinus, 38
Siddha healing practice, 126
Siddha medical texts, 126, 129, 133, 140
Siddhar, Dakshinamurthy, 130
Siddhars, 133
Sillarai Kovai, 15, 127–8
  see also Pasteur, Louis
smallpox, 1, 14, 101
snakes, 111, 127
  irtuhalakkuzhali, 112
  karuvela, 112
  mandala, 111
  moorkhan, 111
  Nagaraka, 111
  rudhiramandali, 112
  snake pills, 136–9
  vellikkettan, 112 (see also fertility)
Socotra aloes, 2
Song dynasty, 62, 66–7, 77, 88
South Africa, 23
South African medicine (see muti)
Southeast Asia, 4, 10, 20, 91
slavery, 18, 21–2, 25, 111
Spary, Emma, 8
Speciale, Fabrizio, 6
Sri Lanka, 6–7, 20
Stern, Philip, 17, 151
  see also Company-State
Strange, Thomas, 137
Subrahmanyam, Sanjay, 3
sugar, 2
Sullivan, John (1740–1789), 170
Sultan, Tipu (1750–1799), 159
syphilis, 1, 133
Syrian sumac, 2

al-Tabari, Ali ibn Rabban (838–870), 37
Tamil coast, 125, 128
Tamil medical manuscripts, 125, 126, 140–1
  see also Siddha medical texts, Siddha medicine
tamira parpam, 135
Taqasim al-'ilal, 56
Tarjaman, Isa (1227–1308), 72, 90
Thabit ibn Quarra (d. 288 AH), 42, 54
thalassotherapy, 7
Thanjavur (Tanjore), 136–7
"The Account of Malabar," 103
"The Tamil Physician," 15, 130–2
  see also Grundler, Johann Ernst
Theyyam (ritual dance), 103
thulasi leaves, 133
al-Tibb al-muluki, 56
Tibb-un-Nabi, al-Tibb al-Nabawi (prophetic medicine), 108–9, 120

Timor, 23
Toussaint, Auguste, 2
toxicology, 127, 129, 143
Tranquebar, 101, 128–9, 134
Trianthema portulacastrum, 132

Udalkuru vannam, 130
　see also Siddhar, Dakshinamurthy
unani tibb, 4, 6, 144
United States, 20, 21
Urdu, 6, 9
Uyūn ul-Anbā' fī Tabaqāt ul-Atibbā,
　("Lives of the Physicians"), 39
　see also Ibn Abi Usaybi'ah

Vaagada chuvadi (palm leaf
　manuscript), 128–9
*Vereenigde Oost-indische
　Compagnie* (VOC), 8, 125
Vitex negundo, 132

Waguda Tschuwadi, 130
Walker, John (1731–1803), 153, 157
Walz, Jonathan R., 11, 25
waqf, 7

White, Charles Nicholas (fl. 1793),
　161, 169
Wirgin, Jan, 83

Xing, Wu, 89

Yalmon, Nur, 114
Yen, Zou (305–240 BC), 89
Yuan dynasty, 62

zam-zam (sacred water from Mecca),
　108
zar (healing cult), 5
Zeigenbalg, Bartholomaus,
　(1682–1719), 15, 101, 108,
　117, 128–30
Zhang Zhongjing (150 AD–219
　AD), 66
　see also *Jin Gui Yao Lue Fang Lun*
Zhengheng, Zhu (1281–1358), 90
Zhufan Zhi, 71
　see also Zhang Zhongjing
Ziming, Chen, 73
Zingiber officinale, 132
Županov, Ines, 8